Science of Melody

Classical Indian System of music "Swar Shastra"
Analysis of its Science and Sensibilities

I0504639

Bhavesh C. Bhagat

Author of *'Science of Rhythm'*

Foreward

PadmaShri Dr. Rajeshwar Acharya

(Prabhavrang)

Ph.D., Padma Shri Honoree, The Government of India

First Edition, April 24, 2020

Publisher:

Universal Pilgrim Productions

Virginia, Varanasi, Vadodara

pilgrimuniverse@gmail.com

Color Paperback ISBN: 9798613260089

Cover Art: Custom-made painting of Kishangarh School of Art commissioned by the Author and created by Shri Shahzad Ali, son of the foremost expert on Kishangarh (Rajasthan, India) school of paintings, late Dr. Faiyaz Ali.

Lord Shri Krishna with Shri Radhe….Divine Rasa Lila….

Rhythm of Mridang (Taal vadya) and Jhaanjh (cymbals) with Sakhis

Forests of Vrindavan, Brij, India

DEDICATION

I bow humbly to Lord Shri Krishna and his eternal Lilas full of "Rasa" encompassing all Universal music in His "Nada Brahma".

This small volume is laid out in English language discussing for the first time, intricacies of science and sensibilities of one of the oldest forms of musical melodic systems (Indian Swara Shastra). It is with deepest love and gratitude dedicated to Lord Shri Krishna and all the musically inclined souls in the universe who persevere in their humble effort to resonate their frequencies in "HIS" "Swara" and "Laya".

Dedication of this effort to Lord Shri Krishna would not be possible without the grace of my spiritual Guru, Vallabh Vedantacharya Shri Shyam Manohar Goswamyji (Shyamubava) of Kishangarh and Mumbai, who has inspired me to sing with him often in Kishangarh and to proceed and progress expediently on the path of seeking and practicing all aspects of music in service to our Lord's Lotus feet.

रसो वै सहः

(The nectar of Infinite Joy of Universe is the Lord himself)

"Venu Naad" in Khinmey Monastery, Tawang, Arunachal Pradesh
Photo by Author

Table of Contents

FOREWARD

(Padma Shri, Dr. Rajeshwar Acharya)

As one embarks on a journey to understand our Indian music, our philosophy, and our cultural heritage, few thoughts that come to my mind, I am sharing here for the reader. This collection of thoughts hopefully will assist in preparing the required background in a curious seeker's mind to appreciate the ancient musical and scientific concepts presented in this much-needed modern volume discussing "Science of Melody". There are many principles in our lives and systems where we try to impose some ideas as "fundamental" ideas and then we get lost in illusions created often due to rigidities of these principles. In any field, whether it's in science, nature, art or music, the idea that everything has to be in some proportion or that there is some forced relativity in all principles is often misleading when we try to impose some things rigidly as "fundamental" (or rather "The only") principles. By being married only to some fixed conceptions, we end up creating more problems than finding solutions. Creativity by its very essence must be flexible to encompass all possibilities. A reference volume such as this is essential to explore all dimensions of our music; spiritual, scientific, practical as well as philosophical.

Self-Expression and Self-Publicity

There is a difference between self-expression and self-publicity. When one listens to a concert or a musician, what is being presented is not music in a subtle sense. Philosophically, what is being presented through the "means" of music is the performer's "ego". Now if the artist has understood the deeper philosophical roots of music, his or her "ego" would have been dissolved in that music and hence they would be presenting the purest form of music that a listener would immediately connect with. However, this is not usually the case. For the most part, when we hear a musician they are presenting their ego covered in the format of their music. They feel good when "their" taan or their "gamak" is judged to be in superlative, often the listener who has spent some money to come and listen does not have the deeper sense of what to expect so in that "blind leading the blind" process the Ego takes the lead and music takes the back seat.

There is nothing wrong with this, as it is a normal progression to evolve from this state to an "Ego Dissolved" stage. To reach this stage deliberate and conscious effort must be made by the practitioner of musical arts to be aware of the guidance presented in a book like this that tries to go deeper into the spiritual and philosophical bearings of our art. That inner drive for enlightenment must be there in both the performer and the listener to realize "what is the true sense of what is being presented and heard?" When "Egoless" music occurs, the distinction between listener and performer is erased they all become one because at the core we are all one "Brahman". This is the objective and performers of the past have achieved this. Even in modern times, there are performers who have gotten closer to this stage but they are far and few between. However, the fact is that even though a listener who does not know about these things has NOT come to listen to the performer's ego, their innate urge is to feel that joy that comes from "self-less" music and that is the Universal music.

Ancient Texts Terms - Sangeet Ratnakar

Now let us consider one of the most referred Indian musical treatises from 12th century AD, Sharangadev's Sangeet Ratnakar. We must understand what is being defined in these important treatises. Word "Definition" means a quality or something which "Defines" an object or a thing. However, there is a difference between "Define" and "Definite". In modern times, we have forgotten these nuances. When we see old treatises, define something we still must ask ourselves is this THE definite answer or description. Always understand that art and human life must evolve continually and maintain their past heritage else, death occurs. Therefore, there is a risk of being hung up only on "Old Definitions". E.g. When one drives a car, the word "Steering" Wheel has another meaning but when you are having a Board meeting and trying to form a "Steering Committee", the same word has a completely different context. Therefore, Sangeet Ratnakar most definitely is the essential starting point for the student of Indian music who wants to study the philosophical genesis of our music, but it should not be the end. The practical and modern thoughts must be married to the ancient reference and creativity must continue onwards.

Precisely using the above examples, we need to understand that in terms of Indian Classical music the terms are often emotionally signifying some deeper sense of a feeling. Moreover, sometimes the terms are symbolically expressing an idea. We need to understand these emotional (bhavatmak) and practical, logical ideas and be able to differentiate between these concepts when we study music and its sensibilities. If a "term" is being used for some idea than over time, it becomes "terminology" but there is also the concept of terms and conditions in studying something to understand it in its wholeness. Practitioners and students of music often misunderstand all these finer points. Without understanding the above nuances, we have created problems when we have forcibly tried to interpret and rigidly force meanings on old definitions in musical contexts.

In Indian "Devanagari", lipi 'Pada' in Sanskrit means 'statuses in English. In addition, the same word 'Pada' has another meaning in Hindi as a poetic verse and it means feet. ***Therefore, the meaning of 'term' is relative based on the context and awareness of the person trying to understand the idea.*** However, if our thought is rigid with forced ideas then we are stuck.

Consider an airplane that is flying above the clouds. For the people inside, things are stationary and for all practical purposes, the plane is not moving. However, someone who sees it from outside and measures it "Relative" to some longitude and latitude only considers the motion of that plane. Then one finds the direction and path forward of that plane.

In our musical approach, we must identify these sensitivities of relativity rather than rigidities of ideas only simply because some 'term' says so in some ancient text. This is the need of the day for our music to continue to evolve. The ancient texts are extremely essential, but they must be studied with an open mind to get their complete holistic understanding in the modern context. The effort of the present author in this regard is highly commendable as this volume is one of the first to combine the experiential sense and sensibilities of our music with their philosophical and scientific evolution.

Dhrupad and it's meaning

Dhrupad has its own meaning but the "Scholars" are quick to simply point it as related to Prabandh Gaan and consider Dhrupad as a composition. This is an example of not considering the true sense and sensibilities of the original meanings and only focusing on terms without any context. When in fact Dhrupad derives from Sanskrit 'Dhruv Pad' meaning the 'highest song'. This Dhruv Pad is the means by which the Providence since our ancient times has established the "Dhruv" (pinnacle) state of music using this format of singing. It is said in our ancient scriptures that the only perfect renderer of this "Dhruv" pad is the Lord himself. Mere mortals can only strive to achieve this perfection and in that objective is our movement and progress of our music. So, Dhrupad is not just a mere composition. To say that is ignoring this deeper sense of its meaning.

Sense of the Indian Swar System

Let us consider some more clarifying ideas from our incomplete and continuously evolving modern understanding of music. In Indian definition, there are Ten Lakshans of characteristics of creating a Raga. However, by reading or writing these ten definitions and copying them 100 times a Raga will NOT be formed. This is the logic of modern teachers and experts who have confined themselves only in the rote limits of 'Defined Raga' and the 'Confined musical creativity'. Just mimicking and reading old definitions will not do the job of truly creating music. For that, we will need to probe our deeper understanding of the sensibilities. Let us explore the Indian Swar system.

Purpose of Aroha and Avroh in the Indian system has been designed for the student to learn to "Create" music not just be confined in its limited definition. An analogy is that when one learns a new language first, each letter is explained and taught and when one learns the whole alphabet, they can write sentences, create new meanings, and understand. This is the purpose of the Indian Swar system to create music not to be confined simply in its definition.

There is a common conceptual understanding that Aroha means to ascend from note 's' and reach higher register 's', and Avroh means vice versa descending. This is true as far as purely a mechanical definition is concerned but this is far from its original intended purpose. In reality, what Aroha and Avroh mean is deeper. Within the seven notes, the note where we start our musical approach is called "Graha" Swar and where we end is "Nyas" Swar. Therefore, the process of traversing from Graha to Nyasa is Aroha and Avroh is reverse. When understood from this perspective Ragas start presenting themselves with new possibilities for a curious seeker. This is why we must look beyond purely mechanical and rigid definitions in something as sensitive as music.

Now let us consider only two swar, s and r. So even with two swar, we have Aroha and Avroh. s to r is Aroha and r to s is Avroh where s is Graha Swar and r is Nyasa swar.

Now add the ideas of some other Lakshans of Raga Sthayi, Antara, Sanchari, Aabhog, Upchar and these add to the Graha and Nyasa and we have a system of seven Swar and the complete way of creating with these seven notes now; s r g m p d n s.

Now we can make any Swar Graha and Nyas and reverse the mechanism. The loop between these two cycles is what is creating the crest and wave formations of melody. E.g. Let us start with 'g' as Graha and 'r' as Nyasa Swar and we can create new possibilities like this for example;

grgmpmgr…gpmgr…gdpmgr….grsrgmpmgrgmpmpgmr

Another loop is being formed now. Now make m as Nyasa (end) and r as Graha (initial) swar. A new loop can be formed as;

Rgmpm…rgmpdpm…rsrgmrmrpmrdpm…This crest and wave formation between Graha and Nyasa are the real Aroha and Avroh. In a purely musical sense, Aroha is the "order" mechanism and Avroh then follows through with the needs of Aroha. The ten Lakshans of Raga are the "Creative Elements" to create the inner feelings of music. They are not rigid definitions alone.

What happens in present times is that the practitioners are neither curious to delve deeper into these finer aspects of creation nor do they often have correct guidance. We find that instead of learning and emphasizing the finer 'sense' of Indian melody, the rigidity of definitions is imposed on the mind and this rote knowledge is then transferred in actual practice. This results in incomplete understanding and rarely a few modern teachers caution or teach the students these deeper aspects. In reality, when a practitioner has with humility made the effort to study and practice these inner sensibilities of music, then a person who has learned only 4 Ragas can perform 400 Ragas as well. With the elements of creativity being the center of focus, one can create Ragas on stage on the spot and small and large compositions can start occurring without effort while the name of that Raga is still unknown because it is 'spontaneous'.

Genesis - Sound, and Space Element

Musical Knowledge is not created in the Universe, it pre-exists. In other words, Sound Element is never produced. "Naad" Sound Element is essentially made available by the Providence in creation itself as a preexisting requirement of creation. However, after much guidance from a learned Guru and committed practice and humble labor, musical Shashtra is self-revealed at some point in time in the true seeker's quest and the music is essentially made available from that point on in its truest form for them.

Now let us understand the two sides of the music. If music were a coin, Sound and Space would be its two sides. If there is water in a bowl, the moment the water moves and is displaced in space, the sound is created. Space is essential for the creation of sound. Without space, there is no sound. This subject is closely related to the Aahat and Anahat Naad that have been discussed in this book as well. The moment the water moves, the "Sound" element makes itself known by leaving its trace that "I was displaced... I have been here already."

Musical Evolution - Dhvani and Gati

Therefore, in this musical evolution what is the starting point one must ask. It is not Taal and Ragas; it is not even Laya and Swara. What concerns us even before these ideas is the concept of audible sound and speed (Dhvani and Gati) came into being from existing Sound Element "Naad" and Space. These are also interrelated as there can be no audible sound without Speed and there can be no Speed without audible sound. How are these two existing elements of the Universe managed to create music? This is where the concept of "Yama or Kaal" in Sanskrit comes into the picture. Yama is the God of Kaal (Time in Indian philosophical thought). The word Yama essentially signifies order and disorder by imposing its limits on any event be it life, nature, music or many other aspects of the Universe. When one claps, the sound produced by the claps is Dhvani, the clap signifies the speed, and it can be varied speed. However by leveraging "Yama" (time) when we enforce a "Niyam" meaning 'systematic control' of the clapping then that same Gati (Speed) is now controlled and transformed into what we call "Laya".

So in essence, first, the Yama becomes "Samyak Yama" or "Saiyam" (controlled act) to create Laya and then through "Niyam" (Quantifiably controlled act) by adding some systematic numerical control over the Gati, we then have created a "Taal". Therefore, the claps can repeat continuously 1,2,3,4 and the loop repeats infinitely. Now the question comes as to how will we know when Sound starts in this clap or when one loop ends and another starts. For this, again Niyam and rules were established that use Heavy clap "taali" to signify starts of the loop and "empty" clap (wave in the air to signify the ending of the loop). In the simplest of forms, this is how Taal came into being as Rhythmic science.

Sthayi Antara Sanchari Aabhog

Now we understood some basic evolution of Sound and Speed. Dhvani and Gati. Let us turn our attention to a deeper analysis of the Structure of the music being sung. Most Indian Classical compositions are structured in the oldest form being practiced today as Dhrupad compositions. These are composed of Sthayi, Sanchari, Aabhog, and Antara.

Sthayi – This word is made up of Stha dhatu in Sanskrit, which is related to a place (Sthan) or a destination. Now in any journey, if there is no destination there is often no objective purpose in that journey. One must know the Destination and then the journey would be worthwhile. Now in any movement or journey, two things are needed one is where you are starting from you must know that and where you are headed. These two points and their relativity to each other determine the fixed portion of the journey. This is the Sthayi "Fixed" aspect of the musical composition signifying the beginning and ending destination and their relation to each other.

As an example, consider a riverbank. You cannot have a riven bank if there is only one side. If you can determine that, you are on one side and that you need to reach the "other" side of the bank then now, you can measure the distance you will have to cross in the river to reach across. This distance is the

"Sthayi" portion. It is fixed because the banks of the river do not move the river moves.

Antara – Once the two sides in our example of the riverbanks have been decided then we need to think about what remains in between. This distance or space between the two banks is called in Hindi "Antar". This is Antara the development of the composition between the two fixed points of reference of the banks of the Sthayi. So the Sthayi is the static fixed form of two ends of the Antara.

Sanchari - Now in this space between the rivers banks the water flows in the middle in varied forms sometimes fast sometimes slow or steady. Based on immediate mood and surrounding elements of that flow such as rocks or islands etc. This creates a unique value of the specific flows in the river. IN our case the river is nothing but the flow of Naad and these varied creative aspects create what we called "Sanchar" of the composition. Development of the music in its creative element with various yet bound in the two banks of Sthayi.

Aabhog – Now these Sanchari, Sthayi and Antara formats all have two participants, singer, and listener. They both play their role and experience and hear this creativity in their own unique ways. The listener slowly understands the variations and the fixed aspects of the compositions and the singer starts feeling more and more his or her creativity evolving in this context. The key element of this is the "Experience" of the creativity by the singer. The experienced value of that emotion of creativity is defined as "Aabhog" in which the singer experiments and presents across the entire vocal spectrum their full emotional state of experience of that composition. Aabhog is as such culmination of the act of singing where the singer is not singing for the audience but gets involved in their own experience of the joy of creating that Naad.

Music as Applied Philosophy

Most performers of modern-day music on stages in bigger cities nowadays know of music as an Art form and try to excel in its physical delivery. However very rare is the practical performer who understands the true nature of music and its relativity to Philosophy. The evolution of musical creativity is grounded in naturally observed scientific and geometrical as well as numerical concepts. However, there is a profound Philosophical element to music as well.

In Indian musical Shastra be it Swar or Taal, the entire idea is based on Trinity. Trinity of Melody, Beats and the Gait/Tempo (Laya) defines music and is essential to the existence of music. This parallels in many philosophical schools of thought, the ideas about the Trinity of purpose. Creation, Destruction, Sustainment, is one example of human philosophical thought where our mind wants to understand the purpose and meanings of these actions in nature. Let's take the "present' moment for example. In musical trinity, we measure the action by the physical act of clapping or some strike to indicate the Tempo and beat of the Melody. The moment we clap, that clap is instantly relegated to the past, what lies in the future is the next beat or next Swar. Then what is present? We may ask. *The present is 'there', it is omnipresent.* Who is singing, who is listening, and where did the performer and the listener come from? These are profound questions to ponder over philosophically.

Often we find learned intellectuals discussing ideas trying to find Philosophy in music, Philosophy in Art, and Philosophy in Aesthetics. In reality, if we just flip flop these ideas we get closer to the essence of the creative spirit even within Philosophy. Rather than finding philosophy in music and Art, one might benefit as well to understand the "Music of Philosophy". The terms are reversed and the pursuit takes a completely new meaning for our sensibilities to explore newer dimensions of both music and philosophy. Wherever there is logic, there must also be some 'Applied' aspect.

The 'applied' aspect of music is nothing but Philosophy. *"Music essentially is thus Applied Philosophy"*. Once we make a humble effort to understand its true depths, pursue it, and leverage it to get closer to our Providence music starts guiding us on the eternal philosophical quest to understand our past, present, and future. This 'Applied' aspect of music needs an understanding of the realms of Aahat and Anahat Naad, discussed at length in the first part of this book 'Science of Rhythm'. This much-needed discussion continues even further in-depth in this volume, 'Science of Melody'. The true artist or the "player" is inside all of us HE is the creator he is the "player". Our physical forms have an expiration date, a limit. Our hands can only strike the Mridang or Tabla so many times. Our instrument of singing, our physical body, and its vocal chords can only sing the material finite number of times in our physical bodies' lifespan. The ongoing process of physical decay takes over at some point. So an artist can study all the musical material in one lifetime but there will NEVER be enough physical time to practice and perform all of it in ONE lifetime. *This musical journey, in essence, is a journey of eternity and continuity.* Therefore, it is very important to study the human genesis of our physical bodies and their connection to music as has been presented here.

Music first occurs in mind and then it is brought out by various psychosomatic actions, as we will explore later. So essentially the artists who understand the importance of Memory singing can bring the "Shadaj" Sa anywhere inside or outside and relate with the whole world in that baseline of Shadaj. This same concept of Memory rhythm applies in "Taal Shashtra". Once you have explored the "Laya" outside by practicing slowly, the effort should be to explore the Laya 'inside' of the artist's soul. Taal also has to be sung just like Swar but it can be sung only by those who have understood the inner Laya. The "Tihai" in Taal is nothing but the concept of the Triad of creation. It will keep penetrating the space and create new loops of rhythm. This fundamental science very few understand. Doyens of Indian music like Pandit Shri Kishan Maharaj and NL Mridangacharya Shri Devakinandan Maharaj understood this. In every performance of theirs, they played the Taal composition three times. When everything is in three wherever you go it will be Tihai and beginning of new creation. Because they knew the secret of logical and applied relativity of music and Philosophy. The musical tempo "Laya" once we fix a certain tempo there is a middle of it. Laya is as much outside and as much inside. So either we reach the middle of the Laya in the one which is manifested outside or we take the Laya inside and experience it inside our souls. Either way, we will find the center. The place of the middle is essential to understand for the practitioners. Those who can master Laya must know not only how to ascend outside but also how to descend inside.

Cosmological Unity of music

Once we understand not only the Universality of music but also the fact that its genesis is in a Cosmological Unity, we now approach the roots of the sound element and its creation. With blessings of profound Gurus, one can gain this knowledge. Everything is essentially ONE. The "Taali" the "Clap" of our hands is THE Taal. In that, Taali is our present moment and it is looking at the past and awaiting the future. In the present cosmological context, the human soul is but one tiny fraction circling the same Sun and center of Energy that other planets in our ("one of many") systems are rotating around. We just happen to be doing so relative to our Earth. Everything in our Soul is Unique yet it is united in this revolution around the Energy. Take the example of a train that is moving with a glass of water across from you on the table in that train. The Earth's gravitational energy is affecting you, you are moving and shaking, yet the glass is not affected by you but the same Earth's energy is affecting water in the glass, the molecules and atoms within that liquid are all connected to each other and they all are affected but the movement. There is a difference between you and glass and the liquid inside, yet all are commonly bound by the uniting factor of the gravitational pull. There are many other natural energies, but the point is that these natural phenomena when you start going deeper or you start going outwards in expansion all become unified. There is a Unity amongst this Diversity of natural laws. Energies transform into each other. Sound affects light and vice versa light affects the sound.

Ancient sages of our Indian heritage knew these facts and modern science is now researching and still documenting.

In music, what we need to understand is this Uniting factor. That must be an easily understood fact. All other complexities can be evolved from this, once we understand this simple Unity model. *All is ONE.* Take Taal, Raag, Laya, and Swar. Remove any of the above components and analyze will music be created without those two. E.g. Let us remove Raga and only have Taal, Laya and Swar will we have music – Yes. If you Remove Taal and keep only Swar, Laya and Raga then also music is possible. However, what if you remove BOTH of the aspects of Laya or Swara then will you have music? NO. However, if you have either one, if there is Laya, then Swars will automatically follow because they are united in their being. Similarly, if you sing in good Swar then automatically Laya will follow. When one who has speed, moves with a constant well-managed speed (Laya) the form of Taal is automatically being created. When one does the same with melody and manages that melody in a well-structured manner, Raga is automatically created. One is the entryway to another because they are all connected with the cosmological Unity we discussed above. To master the Laya a practitioner must not only move forward but traverse the way back to the origin as well. When one realizes these inner sensibilities a practitioner comes to the understanding that the purpose of music is to "Dissolve" the Ego into this Unity not "Evolve" it separately from the natural Unity.

Present State of Indian Classical music

If we are honest with ourselves, we must ask the question, how many people truly want to experience music in its complete Unity. Do we want 'Music' from music or do we want Money from music, or do we want 'Fame' from music? Often you will observe that if we are honest and we introspect as modern professional musicians, today we will find that we want everything BUT music from music. Let us assume that Goddess of music and all knowledge Ma Sarasvati comes directly and blesses a musician. She says I will give you either of the two wishes, chose one.

First – I will give you a boon that you will sing and sing so well that the whole world will realize the beauty of your music and there will be no greater artist in the world than you. You will be famous worldwide but you will NOT be able to sing like that ever again.

Second - I will give you a boon whereby if you sing anytime, Nature will stop. Every molecule and atom in Universe will resonate with your music you will become ONE with them, 'YOU' will vanish but your music will permeate the Universe. The rocks, mountains, rivers, trees, all living beings will sing with you, but 'YOU' will not be seen; your voice will be heard.

Guess which boon every single one of the modern musicians will want - the first one. Such is our state of affairs in modern times. Unless the public claps, the modern artist does not get satisfied. He or she is unable to be satisfied by connecting their music to nature and being one with it. They want material clap, and money now and more money later. Let us be famous, first let us buy a small house, and then let us get bigger, bigger, and bigger. There is no end to greed and bigger. What we need is to go deeper into that music which got us to the stage and understands genuinely in its fullest capacity. Make the effort to dive and depths will reveal themselves to you but today we do not want 'Music' from the music we want everything else. This is unfortunately our present state. The music is not to be blamed in this. The divinity of music showers its grace universally on all. The question is do we have the appropriate vessel to fill ourselves with it and *'share'* it with others. Alternatively, is our vessel mostly full of our ego and our greed? This is the question each modern practitioner of musical arts who is genuinely honest must ask themselves.

The Penultimate (Satik) Swar

What is the perfect pitch, the perfect Swar? When Ustad Bade Gulam Ali Khan Sahab used to sing, some artists used to joke that for heaven's sake, just once even sing one Swar, which is imperfect. That is how perfect his rendering and mastery over each Swar was that it was penultimate in purity. The singing style of Dhrupad in its name itself means "Dhruva" penultimate pad, and what this singing style most requires is the performer to devote their entire soul to identifying and mastering as accurately as human can produce the purest form of "Dhruva" Swar (Penultimate Purity). In reality per the Indian Swar Shastra, humans can only achieve a certain degree of purity to the purest Swar. The only being that can sing and produce the purest sounds are the Creator 'Brahman' himself and hence he is the master of "Dhruva" swars and Dhruva pad. All other singing in the Universe is a shade below his penultimate quality of purity.

What each one of us has as humans is the innate need to share. What we like we want to share with someone else. This is a natural quality of living organisms in many different degrees but it is universal. When a creative artist is studying and practicing to produce the purest form of Swar in their singing and performance, they want to share their joy with someone listening. This is the natural state of pure art and performance. Today this creative spirit is covered up by the "Ego" spirit of the artists more often than not and hence the urge to share and make someone else experience the joy of pure Swar is lost amongst the urge to rapidly extract some "Wah Wah" from the audience or the listener. Music and its purpose are not to "Resolve" the Ego but rather to "Dissolve" it. Only then, one can start the journey towards the Penultimate Pure Swar.

The secret of those musicians whom we call "Greatest" artists is that they often sing or perform for their own self and not for the audience. They lose themselves in their inner joy of navigating the Purity of the Swar. The listeners feel it instinctively. Where this element of "playfulness" is absent music becomes a burden and laborious task to keep a long face and sing. Then everything is artificial. In reality, the "Dhruv" spirit of Purest Swar is preexisting in nature as

26

designed by the Creator. It exists in the winds of the Mountains, in the Hum of Oceans, in the Sweet songs of the "Koyal"....what one needs to do is to take those pure elements and practice those not labor to create anything new. These things already preexist the key is to lose yourself and dissolve your ego in the purity of natural sounds and then one can find the pure notes and swar in their practice.

So what is the Practice? How can one get closer to the Purest Swars? This occurs in Three Stages.

1.　　　　Attention – When singing the Attention is not necessarily on Singing, it is on "Listening" because when you concentrate only on words and your speech, you will lose the frame of reference that is needed to attend to for Purity. This attention is the most fundamental quality required to start getting closer in practice to the Pure Swar. Practically speaking that is why a quiet and isolated corner of a location is preferred for practice. Because each Swar must be explored in its fullest form and the singer must try to achieve the closest resemblance to its reproduction from a Harmonium or Sarangi or Vina.

2.　　　　Concentration – Slowly when you are attentive to the degrees of closeness with the Pure Swar, one will start understanding the moments when we come very close. In those moments, we feel that the Swar we are producing matches the resonance of the instrument accompanying the singing. These moments then automatically force our minds to start concentrating on reproducing such moments. In my earlier book on Science of Rhythm, we explored the nature of Parallel Concentration (Avadhaan). That is the natural leading stage proceeding from Concentration.

3.　　　　Meditation – The more advanced one gets in their Concentration, we achieve Parallel Concentration and then slowly Meditation. This final Stage of Meditation occurs when the performer is detached in their act of singing from the world that exists outside of them and their focus tends to be ONE with the spirit inside. This then becomes the journey of Devotion and the bridge between Anahat and Aahat Naad is built here. In our Universe, there are many stages of Consciousness. Conciseness is not just what modern science can measure it exists in many forms, e.g. Botanical Consciousness amongst the natural world of plants and trees and forests, Animal Consciousness amongst the animals of our planet, Human Consciousness amongst all living population of the world. These all are connected via Supramental consciousness connecting all humans to the ultimate Divine Consciousness of the Providence who created the entire Universe. The Meditative state of music helps in traversing these planes of Consciousness.

Music is Play

It is said that the main work of the Creator is to play, he does his "Leela" in the Universe and all good, bad and ugly ultimately gets encompassed in HIS play. There is no effort for the creator. This element of "Leela Bhava" or playfulness is essential to understanding true nature of music. Music just like Sports is play. If it becomes a burden it is not music just as a child who is playing Sports will not enjoy Sports if Sports becomes academic only. However, to understand Sports one must also understand Sportsman Spirit. This is a commonly used English term and in music, we must equally emphasize this "Spirit". When the "Correct" sense of a term is understood, it helps in building the correct "Character" of that term in one's mind. There is character in music so long as its spirit and execution is Correct. When it is incorrect, the character of music is being lost. To find this "Correctness" in sports of music is the job of the performer or the Sportsman. A "Referee" in sports can only refer but he cannot judge as well as execute the exact timing of your shots so that you achieve your goals. In the same manner, a Guru in music can only refer one to the path of correctness, but how to execute and obtain its Spirit is up to the conscientious practitioner of the musical Sportsman. Character in music has to evolve and that take patience and time. Music just like Sports makes one stringer not just in exterior but also most importantly in interior aspects of the Human existence. However just as we clean our body of internal and external impurities to remain physically healthy, in music one must make daily effort of being cleansed with practice to obtain the correct character of each note. IN that sense, Correctness and Character for musical performer are the same. *"If wealth is lost, nothing is lost. If health is lost, something is lost. If character is lost, everything is lost."*

Categories of musical Performer

We can identify several categories of musical evolution of a performer.

1. **Shikshakar** – The Teacher is the one who can give "Shape" / "Aakar" to the Knowledge "Shiksha" that he or she possesses so that the observer or listener can luckily and rapidly understand the true nature of that knowledge of music.

2. **Ankur** – This is the category of musician who can only "Copy" but has not achieved inner depth to create something new.

3. **Kalakar** – This is the entertainer who entertains with their "Kala" or musical prowess skills. They emphasize showmanship and gimmicks but rarely go to depths of the inner nature of music.

4. **Bhavuk** – This is a musician who is "emotional" or "Bhavuk" they have mastered the inner feelings and senses of music and can share those experiences with the audience. They don't need to rely on Showmanship and rapid gimmicks, he or she can produce immediate joy in listener b touching their senses. Audience responds without any effort with "Wah" and "Aah". This is hallmark of a great performer.

5. **Rishakavastha** – This is the penultimate perfomer. This musician has reached the inner spirit of music and has become one with it. When this occurs they can sing and the audience becomes "The singer" The music encompasses both singer and listener and they become one. Time stops and pauses. This is the ultimate stage.

There are seven notes in all of us whether we are listener or performer. Rather as we discussed earlier there are seven notes in every element of the Universe. A Divine musician knows that the Listener is one of his strings and he will pluck the strings in their heart and resonate with them to become one with them and their music will create a Union. As seekers of true music, we must make an effort to identify these 7 Swar not just in ourselves but in the whole world surrounding us. When we realize how beautiful the Providence has created and provided us with these 7 amazing notes and when they can be sung in its purest form, magic starts to happen. This is the objective of a truly humble seeker.

"Correctness" of musical Notes

The key here is for the beginning musician to try and first identify the primary note "Shadaj" first inside of himself or herself by practice with a Harmonium or another instrument. Study that and concentrate on that repeatedly until your soul identifies its true nature and you will know it when it happens your body and mind will tell you. Then from that correct "Sa" automatically you will feel and produce what we call "AntarGandhar" or "Inner G" in Indian notations. From the two of these third Swar automatically will be produce. For giving birth to a third, the first two are essential. Then you naturally will gravitate to a "Swarit" Madhyam Swar and this all happens internally in the mind, which has practiced "Purity" of the first Swar Shadaj. This Purity consist of four dimensions of any note. Pitch, Timber, Duration, Loudness. If either of these four are not correct than the note will be not Pure. So the dimensions of the note must be understood by the beginning musician and they must strive to produce the notes slowly but perfectly in the beginning. All else will naturally occur and follow.

Inner Nature of music

When we try to learn or practice anything that is as universally available as music, we must make an effort to go to its inner sensibilities. We have created in modern academics blunder after blunder by focusing on the outer subject but ignoring the inner nature of that subject. For example, we have studies in "History of music". This in itself is a wrong starting point, how can something like music, which is Universal and is permeated into every atom since creation of the world be studied for its History. It is precisely for that, reason the present volume discusses the Yogic and Philosophical evolution of human form itself since Music is intricately connected with our evolution.

What we must emphasize is understanding our "History in Music" or Studying of our "History by music" when we flip the context of our approach the hidden meanings become available to us. In same manner, we award degree of "Masters of music". Such a degree means nothing in essence for a true seeker, because how can one "Master" this music in one life when its inner nature is eternity. Rather the focus should be on "Mastery of life by music" When we understand the eternal phenomenon of music and its applied nature in universe we start gaining mastery in many other aspects of life itself. Music is a fundamental phenomenon it is not an artificial act. There is difference between artificial and artistic. When the "natural material" is exhausted then we need to create something artificially. Per the Indian ancient texts 64 "Kala" are described. These are 64 skills of performance of excellence. Of these 59 are materially oriented so that when the natural material is exhausted for these 59 skills artificially new material must be produced. But only 5 of these 64 are called Fine Arts in nature meaning that their "Material" of execution is finer in nature and is creative not artificial and hence it is infinite and eternal. Music is one of these Fine arts.

The emphasis must be on both Fine and Art here because they are sides of the same coin.

Concluding Thoughts

Music is a Fine art that is not to be mastered in its literal sense, but rather experienced and eternally explored with humility for a true seeker. For once one thinks they have 'mastered' the musical art, the artiste then has left the path of continuous growth and is on the path of stagnation. This journey of music is not of one life but rather of lifetimes. When one embarks on a journey of music after analyzing its inner nature, the fear of death and other material fears disappear. This journey must be embarked upon with naturalness discarding all material artificialness. A mother does not need to become an expert in philosophy and all modern science to sing a lullaby to her child - it occurs naturally with the pure ingredient of Love. This is the requirement for study of music. Love and natural persistence with humility.

We must seek to keep our position fixed "Dhruva" in the "Swar". Let us approach this journey in small steps and without any "Pad" or "Title" in our identity. The word "Pad" in Hindi has many different meanings and all relevant here. Pad could mean "Status", or "Title", or "Position", or "Feet", or "Poetry". Our objective in musical journey of eternity is to be devoid of all material status, title or positions and humbly walk small steps with our feet to experience the poetry of natural notes and melodies and rhythms.

Thus, we embark on our journey of "Dhruva Pad". As we discussed earlier. The "Structural" interpretation of the composition is the first part of Dhruva Pad, which is "Sthayi". The second part is the "Antara" which is the other boundary of the structure. Once the two boundaries of the composition are defined, the middle portion that is flowing in between is the third component of "Sanchari". Finally, when one executes upon these three, the fourth is automatically born which is the "Experienced" component of Dhruva Pad and this called "Aabhog". Thus, when one follows these four elements of the music one achieves the "Dhruv" ness in their music. Now this "Dhruv" nature of the melody alone is not sufficient, for the walk to be efficient it must be in some tempo of duration and this is nothing but the Universal Rhythmic behavior, Rhythmic Structures or values. This Melody when performed in this duration and tempo of Universal Rhythm as applicable produces the oneness of Dhruva Pad.

Some words and ideas can be translated but few things are of such fundamental nature in their existence that they cannot be translated. Music is Universal. The seven notes are so pure and profound that when we become one with them, we automatically become pure and the notes, which contain goodness in them, impart in us the same quality. This feeling cannot be translated only experienced. We must not try to mix our mechanism with nature but rather strive to understand the natural mechanisms of music and let it guide us.

Walking, Running, our very Being and every act in Universe has built in natural rhythm try to understand it and experience it and pair it with the seven pure notes and automatically you will have reached one-step closer to your "Dhruva" nature. Let the help come from divine nature. The divine providence is the purest and perfect "Dhruva Pad" creator. The "Dhruvta" or "Superiority" is of eternity. *We must be one with the eternal Providence.*

The Musical traditions that I represent are those that respect our most ancient evolution yet are not afraid to expand upon those ideas and add new creativity in thought and practice. My musical tradition is that of Padmashri Pandit Omkarnath Thakur who did his own research on the field of Musicology and documented and systematically established updated musical theory and practice in the Benaras Gharana of singing. His Shishya Dr. Balvantray Bhatt ("Bhavrang") was my Guru. I have been fortunate at tender age of ten years to embark upon my musical Naad Yatra under guidance of two such learned and devoted souls.

I am glad that efforts initiated by Shri Sharangdev in 12th century still resonate in the 21st century now in souls like Shri Bhavesh Bhagat. He hails from Virginia (USA) and Vadodara (India) and continues this tradition of inner inquiry as to the true meaning and fore-bearings of our musical lineage. I am truly encouraged by such genuine efforts to present our Indian Music to the modern world. I bless and congratulate this fine effort and wish him continued progress in his mission of creating and offering life's music in use of spiritual progress. Even while living so far away from shores of India, his passion demonstrates sincerity and humility to explore the origins of how our Indian musical system and philosophy came into being using Sharangdev's Sangeet Ratnakar as a frame of reference. Observations put forth in this volume by the Author are directly arising from his own unique multi-dimensional perspectives in life that are many to count. This volume to present in English language some of the essential meanings of our Science of Indian Melody would prove to be of immense value as a reference for any serious "Saadhak" of Indian music. I wish all those who open these pages, continued blessings from our Providence to succeed in their individual musical explorations and may they all find their "Dhruva Pad" in their eternal musical journey.

Padmashri Dr. Rajeshwar Acharya ('Prabhavrang')

Benaras, India

February 12, 2020

PREFACE – About this Book

Dear Friend,

If you are reading this preface, then one can safely assume that you have some affinity for music. The purpose of this effort is to connect with music-loving souls throughout the world who want to get to the root of the sources of our Indian musical system. Welcome - present Author hopes this reference volume provides something of value to you as you flip through the pages. In this preface, we cover the WHAT, WHY, WHO and WHERE of this effort and its evolution in the book format.

What is this book about?

This is a second volume continuum of the Nada Yoga series on Indian Classical music and its varied facets of spirituality, sensibilities and scientific and practical evolution. The first volume focused on the 'Science of Rhythm'-Indian "Taal Shashtra" where the Indian system of the rhythmic cycle is explored in depth along with practical ideas covering scientific and spiritual aspects of these rhythmic cycles in Indian music. The 'Laya' is the foundation of all life itself. As such in the first volume, we explored the foundation upon which to build continuously evolving and creative musical structures.

Now, this present volume 'Science of Melody' continues the journey to build upon the foundation of 'Laya', the melody 'Swara' and its creativity of Naad. Earlier volume explored the ten 'Prana' (life force) of Taal. This volume explores the ten 'Prana' of 'Swara' or the life force of our Melody.

Let us for a moment consider music as a whole and consider it from the perspective of the Asian concept of Yin and Yang or Indian concept of Shiv and Shakti. If Rhythm is the Yin then Melody is the Yang; if Rhythm is Shiva then Melody is Shakti. One cannot exists without the other. Hence, in this second Nada Yoga volume, we now cover the depths of the Indian musical melody.

We explore the "Science of Melody" - Indian "Swar Shashtra", its sense and sensibilities as described in one the most well-known ancient treatise from 12th century AD "Sangeet Ratnakar" of Shri Sharangdev. Contemporary viewpoints expressed in the foreword by Padmashri Dr. Rajeshwar Acharya (Prabhavrang) lay the ground perfectly for us to embark on a most joyous exploration of this infinite subject of Indian Melody. The book explores the Author's evolving philosophical, spiritual and practical scientific analysis and experimentations with these concepts as a lifetime student of music from the learned Gurus acknowledged herein.

Our purpose in this volume is concentrated solely on the 'Melodic science' and its treatment as described by Sharangdev in 1200 AD as such, this book concentrates its scope on the aspects of Sangeet Ratnakar, which presents entire evolution of Indian Melody. The aspects of Tala Shashtra are covered in the present Author's first volume titled 'Science of Rhythm'. Aspects of Raga documentation of the Indian system have been compiled by many great musicologists of past such Dr. Bhatkhande, Pandit Omkarnath Thakur (Founding Guru of Dr. Acharya's Benaras tradition of singing), and others. As such, our purpose here is not to go into Raga structures and their well-known practical details but rather understand 'How' the melody of these Ragas came into being. What are the principles behind these melodic systems? As one masters the 'Science of Melody' discussed in this book, possibilities open up for the soul to create infinite new Ragas rather than being bound by the ones that are already documented. This is the true objective of studying the 'Swara Shastra' before studying the Ragas.

Why is this volume written?

There are many reasons to write this volume. Most importantly, if even a single soul gains some inspiration and knowledge, using the words written here in the practice of their musical Science and Art towards a Spiritual goal, then the author's effort would have been worthwhile. "Saarthak"

The subject of Indian Music has been covered over the millennia in many aspects but often the literature is very difficult for a practitioner to refer to and read in a lucid manner. While the practical aspects of Raga and Tala are discussed in such literature often the questions of 'Why' and 'Where' of music are ignored. Modern musical literature, for the most part, can be put in two camps. Students of Ph.D. in Musicology write their doctoral thesis published as a book, but it is often difficult to read, as their objective is too academic primarily and not writing something for the larger audience. The second camp of musical literature is the practical "How To" category where the emphasis is on learning how to play or sing for beginners and the deeper evolutionary aspects of music are not discussed there.

The subject of music is infinite and hence no book can do justice to its depths. Yet due to the above reasons, we often find very technical Ph.D. thesis or very basic beginner introductions to Raga and Taal. Emphasis is often on the 'How' alone without any analysis of the source of the music or exploration of the "From Where" and "Why" aspects of our musical evolution. Amongst all this, the sensibilities of philosophical and spiritual dimensions of music are seldom documented, discussed and written about.

The present volume is an effort by the Author to share such perspectives and dimensions of Music that not holistically discussed in the English language while also covering the ancient genesis of the Indian system of melody. As a practicing student of music, the present Author has several frames of perspective in approaching music. First, is the spiritual aspect of music, second is the philosophical underpinning of the systems of melody and finally third is the practical scientific and physical assessment of how music is 'extracted' by the practitioner from the universe and deployed in practice for their own as well as others' enjoyment.

Being an Entrepreneur, Electronics Engineer, Martial Artist, an avid student of Philosophy and a Musician, this is a humble learning effort of this Author to explore the 'Swara Shashtra' holistically in its entirety of existence and evolution. To build a bridge of music not just for joy but also between many lives of the past and the ones yet to come. Music is a thread that unites the human evolution in all our pursuits of life and a true seeker must approach it from all angles to understand its inner meanings and depths. Below also is shared some additional background that applies here as well and was covered in 'Science of Rhythm' as to why these books are necessary for the modern context.

The present state of Indian Classical music is not in the original form that the founding musical Sages envisioned it to be. However, just like life itself, Indian Classical music has evolved and its purpose and its practitioners have evolved. This is good in a way because if art does not evolve it slowly withers away. As Dr. Rajeshwar Acharya points out in his foreword, nowadays the sincere thirst to go to the "root" of all music is often missing in the practitioners and listeners both. Our effort here is to get at the "root" of our 'Swara" Melody and its structures.

The "Spiritual" practitioner who takes the challenge to focus on the inner world of music, certainly also exists but is hard to come by, and for folks like this author, living far from Indian shores, this is even rarer. The performances given on stage in big cities offer music, but this is often superficial and more about a show of physical prowess for the most part. The performer is trying in the back of their mind, to think of how to obtain another performance, how to get more people to clap, or how he/she might look on social media. Some of this is necessitated by our modern society and is essential even for survival; however, what is truly missing is that "thirst" for understating the "roots".

The Student-Teacher, that is "Guru – Shishya" parampara (tradition) of passing down traditional cultural assets such as music, also has been hampered with vagaries of modern society, ignorance and often materialistic mindset of the teacher as well as the student. Often, the teacher does not teach the full depth of the subject leaving the inner teachings for his progeny (son or daughter) to keep their material source of income going. And many of today's students also are driven by the urge in our world of instant social media selfies, to become *"famous first and experts later"*, ignoring the patience and sacrifices required to hone a passionate art such as Indian Rhythm or any form of classical music.

Often the commitment and persistence in the Guru and Shishya are lacking today. Mostly, because the end goal is not the "Spirit", it is not the "Devotion" but it is rather "Ego". This is beyond our control; this is the reflection of our state of society and its evolution. Being "professional" in something as important as music is a very commendable and most desirable thing in life to achieve. *However if the "professional" is at the expense of the "spiritual" and "emotional" aspects of music and its educational journey, then something is misplaced in that experience.* Music taught or performed with the right "spirit" and purpose can be felt by the "attuned" soul. This is rare. Such spiritually inclined "Naad Yoga" teachers do exist in today's time and we wish them all the strength from Lord in continuing to pass on their spiritual music to their deserved students.

Indian music has two components, "Swara" and "Laya". The business of Swara is conducted in non-rhythmic instruments e.g. "Vina Vadan" and vocal "Gaan", while the practical execution of Laya is conducted through rhythmic instruments and "Taal". Both these aspects of Indian music have lost their original spiritual essence in modern mass performances with an emphasis mostly on the physical excellence of the player. The audience, for the most part, does not want to invest their time to learn about the nuances and depths of Swara and Taal. Such an audience "entertains the performer's ego" with fleeting claps and "Wah-wahs" and the show goes on.

The question for the true seeker then is to find out how to progress in their musical journey in these challenging but constantly evolving modern times. The big challenge is how to appreciate the **original thoughts behind the meanings** of the minutest details and "Bhava" emotions of nature that our ancient Sages documented when they created the Rhythmic and Melodic structures of Taal and Swara compositions.

Who is this book for?

This book is an effort to present a holistic point of reference in the English language for the 'Swara Shastra', Indian 'Science of Melody' and its intricacies. This volume strives to provide a meaningful reference to a musician, practitioner or a student who is already playing some form of Indian melodic instrument or Indian classical singing. An "Active" listener who wants to make genuine efforts to grow their understanding of what is being performed in melody and what their ears are hearing and their minds comprehend would also gain immensely from some of the ideas here. The book can be used as a reference volume on Melody to refer time and again as one's musical evolution proceeds.

The book expands upon the Science (logical and experimental roots) and Sensibilities (emotional and spiritual feeling based genesis) of the Indian Melodic structures also known as "Swar Shastra" in Sanskrit. Like all forms of Science, some preexisting background and conditions have to exist for a seeker to understand the postulated theories and the experiments. The Science of Melody is no different. To benefit from understanding the ideas and nomenclature in the book one must meet some of the following requirements;

1. One must be curious to explore and be willing to comprehend the 'Source'; the depths of Melody

and its structures beyond the physical aspect of singing or playing the instrument.

2.   One must be familiar with some basics of singing or playing a melodic Indian instrument. This book is NOT a guide on "How to Play" the instruments or "How to Sing". That subject is well covered by multitudes of books available in Indian Ragas and their compositions. Rather than discussing practical singing instructions of an individual Raga, here we are exploring the roots of the evolution of 'All' Ragas. These fundamental concepts are to be 'experienced' by practically singing and/or playing an instrument preferably under the guidance of a Guru well versed in Swar Shastra.

3.   If no such Guru is found, the seeker is to continue self-progressing and self-learning the physical aspects of the Singing or their chosen Melodic instrument by "Active" listening and practicing without fear. By making an effort to find and learn from some of the written materials as this Author has found. When the right time comes, 'Guru' will present himself or herself, what is essential is the commitment, and there can be no better proof of truth in this eternal secret principle than the life of present Author itself. A true seeker of music is fearless.

4.   The reader must have a spiritual and emotional bent to appreciate the dimensions of music that cannot be described in a book but rather can only be experienced in solo meditative practice.

5.   The reader must be willing to experiment with the ideas. Even a single basic point once understood must be connected with his/her own personal practical musical experience and seek the guidance of a Guru who will assist in continuing this wonderful exploration further.

6.   While the knowledge of Sanskrit and/or Hindi is not essential for 95% of the contents, some basic curiosity about Hindi words and their meanings would only further enhance the Reader's understanding.

How is the book structured?

The author writes the initial Scientific and Spiritual connections of Swara and Naad Brahma from his continuously evolving practice and studies of philosophy and music. All Swara Shashtra sections representing Sanskrit and Hindi pieces of the present volume are compiled from aspects of the 12ᵗʰ Century AD treatise 'Sangeet Ratnakar' of Shri Sharangdev translated in English by late Dr. R.K. Shringy in 1977. The explanations have been written by the Author combined with his own personal experimentation and evolution of music as well as from instructions received from Author's Gurus. Gurus who have blessed the author with their kindness and knowledge and inspired him to pursue this pilgrimage of Naad (Swara and Laya) are acknowledged in the "Bhavanjali" Acknowledgement section.

1.   The section titled 'Science of Melody' is a cross-referencing section between both previous volume ('Science of Rhythm') and this present volume and lays out the oneness of all Musical objectives from the point of view of Science, Religion, Art and Philosophy from a treatise on 'Rasa Drishti' (Perspectives on Rasa) from author's spiritual Guru Vallabh Vedantacharya Shri Shyam Manohar Goswamyji.

2.   Section 1 - Sense of Indian Musical System – covers the entire spectrum of philosophical, spiritual, metaphysical and psychosomatic aspects of the

evolution of Music and its relation to our body and soul as well as the creator. This section is based on Sangeet Ratnakar with perspectives from Author's own experiences in Yoga, Pranayama and Martial Arts. This spiritual and emotional perspective on the Indian musical system is lacking from most modern musical texts and hence Author is documenting these ideas for the soul wanting to dive deeper into the 'spirit' of music.

3.   Section 2 – deals with the first of the three Pranas of Swara Shashtra. Of the ten key characteristics (Prana) of Swara Shastra (Ten Life forces of the Indian Melody) Naad, Shruti and Swar are discussed in depth here with Sangeet Ratnakar as the base reference model. Here we dive into the logical, numerical and scientific aspects of the evolution of the Melodic systems.

4.   Section 3 - Deals with the next set of three Pranas of Swara Shashtra. Gram, Murchana-Krama, and Taan are discussed in depth here with Sangeet Ratnakar as the base reference model. Having established the elements of notes and tones in section 1, those ideas are now logically grouped into numerical permutations and combinations to form melodic structural foundations.

5.   Section 4 – Deals with the final four Pranas of Swara Shashtra. Sadharan, Varnalankara, Jati and Giti are discussed in depth here with Sangeet Ratnakar as the base reference model. These sections essentially build the practical aspects of the Indian melodic system that is built systematically upon the first concept of Naad.

This is a body of work that is continuously evolving as the author and his journey of Naad Yatra "Pilgrimage" evolve. The ideas in sections of the book on the Swara Shastra from the old ancient text 'Sangeet Ratnakar' will remain firm in their historical definitions and will remain a steady reference for the future. However, the application of scientific principles and emotional ("Bhavatmak") sensibilities of practical use of Swara and Taal Shastra from a spiritual perspective will continue to evolve. As the author enhances further understandings from his experimentation and from Gurus' blessings, future editions will be revised accordingly. Hope is that this small volume will act as a useful reference for practitioners and listeners alike. This effort is for those souls worldwide who are thirsty for a deeper understanding of Indian Classical systems of music.

When was the reference treatise Sangeet Ratnakar written?

The authors of the old rare treatise on Indian Music and practice were not just musicians and Swara and Taal experts, but above all devotionally inclined souls who undertook their writing efforts for future generations and in service of their God. The present volume is based on an English synopsis of Sharangdev's 'Sangeet Ratnakar' written in 12th century AD. There are older musical treatises in the Indian system as well and sage Bharata's 'Natya Shashtra' is one of the most ancient documented treatises. However, all the ancient theories had to evolve, they all crystallized and solidified to form a single uniform system of Indian classical music by the 12th century AD, and this was the period when Sharangdev documented this more evolved system in his treatise.

Sharangdev and his efforts fascinated the Author not just because of his Musical knowledge but because Sharangdev also had a multi-dimensional perspective of

Music as the present Author. Sharangdev was an accomplished financial professional working as an auditor, he was an accomplished doctor and master of Indian Ayurveda (Indian system of medicine) and he was a devotee of Lord Shiva thus grasping the spiritual aspects of music. All these aspects of Sharangdev's personality from 12th century AD make his Sangeet Ratnakar treatise that much more insightful.

There have been very few efforts in the English language to present this ancient text for the modern English reader. One was in 1945, second was in 1977 by late Dr. R.K. Shringy published also from Benaras by Motilal Banarasidas publisher. His careful effort in translating the entire treatise has been a great source of inspiration for this present study and effort of the Author. Dr. Shringy's 1977 edition has formed the base of reference with regards to Sangeet Ratnakar's contents.

These melodic concepts are the core of Indian musical genesis. In contemporary times, much creative and evolutionary work goes on in the Ragas and other aspects of the Indian Melody system. However, the basic tenets and the core 10 Prana of the science and evolution of Indian Swar Shastra remain as the primary foundation for all music, as they are steeped in the philosophical roots of the Vedas and the ancient Indian scriptures.

There is a dearth of authentic English literature that approaches music and its roots from holistic dimensions of spirituality, science, and logical and practical sensibilities and their meanings. The modern student is more inclined to learn first by reading in English then proceed in other avenues; as such, this present effort to publish in English is essential. This author is joyously

48

continuing the above tradition of humble souls like Dr. R.K Shringy, by sharing their devotion to Indian Music by translating some of the ancient knowledge. Author shares here musical sensibilities and emotional ideas obtained from his Gurus and experiences from his own experiments and practice of Indian Music for Western and Indian audiences. This effort is imperfect, as Hindi and Sanskrit cannot be translated accurately in English, but it is a beginning. All possible imperfections are authors alone and not of the original authors and Gurus who shared their knowledge in Sanskrit and in Hindi.

Sunrise. Tulsi Ghat. River Ganges - Photo by Author

Where did the idea of this book transpire?

The present author has been listening to the various forms of Indian Classical music for many decades and has had the fortune of not only persisting in this effort but also to graduate from being a passive listener to being a continuing lifelong practicing and experimenting student of Indian Classical music in its devotional form. Challenges along this path have been numerous but the efforts have been worthwhile for the soul's progress. Having been born in Mumbai, India, the author has spent over half of his life now in Northern Virginia near Washington DC in the USA. This distance from the root of Indian Classical music has given a renewed vigor and persistence in the author's musical journey with all the challenges that geographical separation brings.

This author has been fortunate as he grew up in India, honed his interest in reading old texts, and over the years attained the ability to read in Hindi, Sanskrit, Brij Bhasha, Gujarati and Marathi languages. This, coupled with the blessings of many spiritually and musically oriented Gurus, has allowed the author to collect a library of some out of print and rare manuscripts on the arts of Indian music. While some of today's modern music teachers often incorrectly guide the students by telling them to ignore written instructions and only to listen and play, etc., these are just half-baked truths and could hamper the progress of a true seeker.

Of course, memorization of "mind and the muscle" is the fundamental aspect of the Indian Classical musical system. But as stated earlier, writing down instructions and reading and learning are equally important as listening and practicing. Writing and sharing one's written knowledge is one of the key ingredients of retaining and expanding one's knowledge of any subject. Each soul is unique in its own learning, evolution, and pursuit of this science and art of music. One must find the right mix of guidance to persevere and keep progressing if one is committed.

Author's ongoing journey *starting as an avid listener of* *"Swara" Melody and Rhythm "Taal"...and playing* *Mridang...Singing... to writing this book ...is connected* *with Benaras and Temples of Brijbhoomi and* *Rajasthan, India.* **Benaras (Varanasi) is the author's musical home, while Kishangarh (Rajasthan) is his Spiritual home.** Benaras, per Indian historic evolution, is said to be the oldest city in the world and the center of Indian philosophical learning since the early millennia and founding of the Indian cultural origins.

For over a decade now, this author has been regularly visiting this holy city to listen, learn and continue his hunger to play the Mridang and learn more about the science of Swara and Taal. This body of the author had no foresight to choose Benaras as his destination, it is actually the city and river Ganges *(on whose blessed Ghats the whole ethos of Indian life can be seen played out daily)* who chose this author to pursue his Laya in the holy environs.

This book is for the souls who gravitate from all over the world, attracted to Indian music. If this small reference volume gives such worthy seekers some essential flavor on the origins of Swara Shashtra and the Melodic science and its format, the author's effort would be worthwhile. Music is indeed a universal language.

समर्पण ।

श्रीकृष्णार्पणमस्तु ।

इस पुस्तक को भगवान श्रीकृष्णचन्द्र के चरण में अर्पण करता हूँ ।

Temple Verandah in Kishangarh, Rajasthan

Acknowledgements
Bhavanjali

The contents and every word and feeling in these pages are not this author's but rather inspired and acknowledged as the blessings of great souls mentioned in this "Bhavanjali" section. *As such, this author lists one of his Spiritual and musical Guru NL Shri Mukundray Goswamyji as his co-author in dedicating this effort at feet of Lord Shri Krishna.* What is being written and being published is what the author has gained from his blessed Gurus and is sharing with all as he continues to learn.

Author wishes to thank the blessings of his Mother and Father along with other elders who have planted the seeds of this joyous journey in his life. Also it would be remiss if the author did not thank the gracious souls who have helped with their various efforts to proofread the manuscript and assisted in some way or other in creating a volume with hopefully minimal errors. Author's daughter Pushti Bhagat who is a student of Western Flute and is also learning Indian Bansuri as well son Brij Bhagat who is a passionately motivated student of Western classical violin; both for encouraging lovingly these publishing efforts. Mr. Sushant Karmarkar and his father Shishirji both residing near Shanti Niketan environs in West Bengal for their help in creating a custom instrument for the author. Both are also accomplished Dhrupad singers in their own right and Shishya of Dr. Rajeshwar Acharyaji who has graciously written Forward for this book. Pandit Badrinarayanji and "Dada" Arun Chatterjee of "Gurukul" music school at TulsiGhat, Benaras who always are eager to spread the joy of Laya with Shri Mahantji. I wish that the Lord grants all these souls their continued success in the pilgrimage of music.

All errors, omissions and shortcomings are author's alone in writing and compiling this work. The mind and our bodies are full of imperfections. This pilgrim asks for grace of the great gurus illustrated here for their continued blessings onwards in this "Naad Yatra".

Author's Spiritual Guru

Vallabh Vedantacharya Shri Shyam Manohar Goswamyji

Shri Shyam Manohar Goswamy (Shyamubava) of Kishangarh, Rajasthan and Vile Parle, Mumbai

Author receiving blessings of his Diksha Guru

Shri Shyamubava represents the 16th Generation in the lineage of Mahaprabhu Shri Vallabhacharya (the founder of Shudh Advaita Vedant philosophical school of Hinduism). He is the foremost scholar and a great literary author in the world in the philosophy of Pure Non-Dualism with worship of Lord Shri Krishna at its heart. He has single-mindedly dedicated his life towards planting the seeds of Krishna Bhakti in those souls who desire to experience the universe with Lord Krishna as its sustaining point of reference. Even in his late seventies, Shri Shyamubava is tirelessly committed to research, author and publish all identifiable and yet unpublished works of Mahaprabhu Shri Vallabhacharya.

He has also constructed a structured philosophical study curriculum and teaches an advanced course in English, Sanskrit and Hindi on "Vallabh Vedant" in the respected University of Mumbai's Philosophy Department as the penultimate expert in this Vallabh Vedant philosophy. **Shri Shyamubava has lived his life by example, as a perfect embodiment and union of all the Yogas.** For the person who wants to understand more about the entire life and his philosophical and literary body of work, the wiki link is published here. For the curious seeker, this will act as a great source of information. https://en.wikipedia.org/wiki/Shyam_Manohar_Goswami

Shri Shyamubava with his divine blessings has propelled this author on the path of continuous learning and listening to "Inner" music by daily self-practice. *Whatever knowledge, spiritual and musical qualities are evolving in this present body of this author are due to his Guru's grace and due to singing with Shri Shyamubava in his home temple at Kishangarh.* Many unknown and "Aprachalit" Ragas he makes up on the fly with his creative devotion. His musical depth is immeasurable and self-inspired by Lord Shri Krishna.

Shri Shyamubava plays Xylophone as his instrument of choice and is well versed in every aspect of the deepest Rasa of Indian music. Music for him has its sole purpose as an offering of a loving ingredient to serve Lord Shri Krishna.

Shri Shyamubava above with his affectionate meeting on the 89th Birthday of Pandit Jasrajji.

Over the years, Shri Shyamubava has also been kind to highlight and illustrate for Pandit Jasrajji, his blessed insights on the intricacies of Bhakti centered devotional music of Pushti Sampradaya. Both Pandit Jasrajji and Shri Shyamubava share a great deal of affection for each other and Shri Shyamubava loves to listen to Shri Jasrajji sing when he is in Mumbai. Shri Shyamubava splits his time in his two seats of residence one in Mumbai at Vile Parle and other at Kishangarh, Rajasthan where he invariably spends the Diwali (Indian New Year) sharing his insights with Krishna Rasik souls there.

Even a brief moment in the company of Shri Shyamubava puts one's soul on the path towards partaking Rasa in Shri Krishna's eternal Lilas.

NL Shri Mukundray Goswamyji, Mumbai

"Veena Vadan Tatvagnya", Bada Mandir, Bhuleshwar, Mumbai

Playing his Veena above and accompanying friend Ustad Amir Khan

Shri Mukundrayji was the uncle of this author's spiritual Guru Shri Shyamubava. He represented the 15th generation lineage of Shri Vallabhacharya. His Grandfather Shri Jivanji Maharaj was a great sitar player of India having learned from Senia Gharana and the famous Sangeet Martand Prof. Bhatkhande was a music disciple of Shri Jivanji Maharaj of Bada Mandir in Mumbai. (Shri Bhatkhande performed invaluable service to Indian Classical music by formalizing and structuring the Ragas in a curriculum format in his Sangeet Shashtra volumes.) Mukundrayji was born into such a musically and spiritually inspiring environment.

Shri Mukundrayji had also spiritually blessed and initiated the author when he was in his early childhood and graced him with his musical blessings to progress towards his path. He realized that the devotional Krishna focused Pushti Sangeet (centered primarily on Dhrupad style of singing) must be formalized and documented as a reference for future seekers. Shri Mukundrayji published the *"Nada Rasa"* volume of Dhrupad and Khayal based compositions of Pushti Margiya Krishna Devotion Padas. The present author has the first edition published over four decades ago of this Nada Rasa volume. This "Bhava" emotion-laden collection of Swara compositions is the author's sole entry point and inspiration for his Vocal and Instrumental musical sojourns. Nitya Leelasth (NL) Shri Mukundrayji played his Veena in Lord Shri Krishna's Seva and his knowledge of music was self-inspired by Lord Krishna.

He was well versed in all forms of classical arts and was known in the Hindi movie industry as well. *Shri Mukundrayji was a great friend and admirer of the* *greatest **Indian Classical music vocalist, Ustad Amir Khan.*** He was friends with Naushad and other creative personalities of the time.

Author's Mridang Vidya Guru

Shri Mahantji Vishwambharnath Mishra

Shri Mahantji of Sankat Mochan Mandir, Tulsi Ghat, Benaras (Varanasi)

Shri Mahantji hails from a great-learned family of Benaras steeped in spiritual as well as the musical heritage of the ancient city. He is the current lineage holder of the ancient 400 plus-year-old "Akhara" and temples and institutions established by great Bhakti Saint Shri Tulsidasji who wrote "Shri Ram Charit Manas" in chopai chand. The Hindu Bhakti tradition is indebted forever to Saint Shri Tulsidasji who wrote Bhasha poetry of profound excellence with Shri Rama as his center point of devotion.

Shri Mahantji presides over the famous Shri Sankat Mochan Hanumanji Temple along with the main seat of Tulsidasji at Tulsi Ghat in Benaras. He is a Ph.D. Electronics Engineer by profession and also is the Head of Department at the Benaras Hindu University (IIT) Electronics Engineering department. He also is Chairman of the Clean Ganga Foundation which is a nonprofit focused on ensuring the River Ganges and its holy waters are cleaned of all physical and biological impurities for future generations to come. *Suffice it to say that Shri Mahantji is the essence of a true Karma Yogi* and this author personally has seen him start his day at Sunrise with Maa Ganga and end his sometimes 18 hours or longer day with Shri Sankat Mochan Hanumanji without fail as his spiritual commitment- no matter how hectic or busy the other material duties.

Shri Mahantji learned Mridang (Pakhawaj) from his Grandfather "Babaji" Shri Amarnath Mishra. He continues the musical Benaras Tulsi Ghat tradition of Mridang playing with a primary spiritual focus. Shri Mahantji's family has done a great service of the Indian Cultural musical heritage by organizing and hosting two of the most pre-eminent international Indian Classical musical conferences since many decades giving encouragement to the arts. These concerts have created a much-needed forum for students and performers who have reached their pinnacle to come into contact with each other amongst a most spiritual environment. Musicians from all over India perform in these two concert series more as a form of their devotional offering to the Lord then as a professional duty.

The *Annual Sankat Mochan Hanuman Sangeet Samaroh is a weeklong all-night concert coinciding with the Hanuman Jayanti festival hosted on the premises of Sankat Mochan Hanuman temple.* This concert series is focused primarily on all forms of Indian classical music varieties except for Dhrupad (for which there is an entirely separate concert). This usually occurs time the month of April/May yearly.

The second series is the *International Dhrupad Festival which coincides with the Shivratri festival in India and occurs usually in Feb/March timeframe.* This is a four day all night long series of Dhrupad performances from artists from all around the world on the banks of Ganges at Tulsi Ghat. This series solely focuses on all forms of Dhrupad variety of Indian Classical music and is accompanied by Mridang (Pakhawaj) rhythmic structures. The family of Shri Mahantji has been conducting this praiseworthy service of Indian music for many decades and this effort is akin to performing "Naad Yagna" in service of the Lord. The present Author and his musical journey are connected in several ways with these musical and cultural activities supported by Shri Mahantji.

Amongst all of the above duties, Shri Mahantji also finds the time with his kind grace and smile to persevere and make sincere efforts to continuously hone and refine his Mridang Vadan skills. This Mridang Vidya steeped in the Bhakti tradition of Tulsi Ghat and Gopal Mandir, he lovingly teaches and inspires the present Author from the USA and his other Shishya Shri Tetsuya Kaneko from Japan.

Present Author had been visiting Tulsi Ghat since 2005, to listen and learn at the Dhrupad Mela in Shri Mahantji's garden. Yet for some reason an opportunity never arose to meet personally or speak to Shri Mahantji. *Twelve Long years passed like the Twelve Matras of Chautaal, then the "Sama" came* and Shri Mahantji and the author met actually for the first time at the same Tulsi Ghat in 2017. He said he had been waiting. This author had been evolving in his own way for the past twelve years with more eagerness... and now the joyous Laya pilgrimage has been initiated.

Shri Mahantji has been ever ready to shower his love and blessings and guide the author's "Thapiya". In one of the meetings, Mahantji blessed us with his insight when the author was little restless knowing that many years have passed and there is way too much to learn and not enough time. He said with his ever-present smile ***"This journey is not meant for one Life....It is a journey of many lives."*** This instantly gave this author the patience and confidence to move forward with his Mridang Vadan experiments with renewed vigor.

(Saint Shri Tulsidasji was a contemporary of another great Saint Shri Surdasji who in the same era was conquering the literary heights with his loving Bhakti poetry in Brij Bhasha with Shri Krishna as his point of focus in the Pushti Sampradaya). The present author has been blessed to have the opportunity to have been in deep contact with both the Ram and Krishna Bhakti traditions.

Mahantji Sw. Shri Amarnathji Mishra (Babaji)

Grandfather of Present Mahantji, Sankat Mochan Hanuman, Tulsi Ghat, Benaras (Varanasi)

"Baba" Amarnath Mishra of Sankat Mochan Hanuman Mandir was the founding father of the spiritual Tulsi Ghat Benaras Gharana of Mridang Vadan. He was a great spiritual soul always wearing a big smile on his face as can be seen in the picture above with his joyous

"Thapiya". Not only was he a musician but also a wrestler and avid supporter of old traditions of the Tulsidasji Akhara dating back more than 400 years. Babaji started and initiated the "Naad Yagna" of International Dhrupad Mela with his friends Veena Vadak Sw. Dr. Shri Lalmani Mishra and Padmashri Dr. Rajeshwar Acharya (Shishya parampara of Pandit Omkarnath Thakur). They were worried that in the winds of Khayal and modern influences of the Indian Classical music evolution, the oldest traditions of Dhrupad Ang would be lost. If Dhrupad artists were not supported and promoted by Babaji and such devoted souls, a great cultural asset could have been lost by India. Dhrupad Ang is more meditative and emphasizes the Vilambit Alap as well as Mridang Vadan due to its spiritual roots. Till this day in temples of Brij Bhoomi and many more across India the devotional music is based on this Dhrupad form of singing.

Babaji revived the old art of Dhrupad music by establishing the original roots of the International Dhrupad Mela in 1975. This auspicious "Naad Yagna" that Shri Babaji initiated is going on for almost five decades now. We wish that this effort continues forever in service of our Lord and assists the seeking souls in connecting with the Naad Brahman.

In those early days, Mridang Vadan was in even more dire conditions. Now after almost five decades of this wonderful committed effort. Mridang Vidya has revived to such an extent that the present Author born around the same time as Dhrupad Mela, is writing a book based on his experiments in Mridang Vadan and Taal Science …residing in the USA. All of this is possible due to the blessings of spirited souls like Babaji and their love for Indian traditions and art with a devotional perspective.

Due to his spiritual role being the Mahantji of Shri Sankat Mochan Hanuman temples and Tulsi Ghat traditions, *Babaji had the music of the "Hanumant Mat" in his veins. He also had an inner urge to play the Mridang primarily with a spiritual focus. He found such a spiritually oriented teacher in Shri Mannuji. Babaji learned Mridang Vadan from Sw. Shri Mannuji Mridangacharya of Gopal Mandir, Benaras.* Sw. Shri Shrikant Mishra and Shri Manik Munde are some of the illustrious Mridang vadaks further continuing Babaji's Mridang traditions.

Padmashri Dr. Rajeshwar Acharya
"Prabhavrang"

The author is having the pleasure of knowing Padmashri Dr. Rajeshwar Acharyaji for over a decade now in the auspices of the International Dhrupad Festival co-founded by Acharyaji almost 50 years ago. Dr. Acharya has been most kind in spending his valuable time instructing this author and discussing his thoughts and his ideas about music. The author is profoundly thankful and humbled by the valuable insights presented as part of the Foreward by Dr. Acharya for this book. The knowledge of Swar and Taal obtained by this Author from a Guru like Acharyaji is priceless. Acharyaji is a multi-faceted personality who at the age of ten years obtained Indian musical training under the guidance of Gwalior Gharana's two doyens of Indian music. Pandit Omkarnath Thakur and his disciple Padmashri Pandit Balvantrayji Bhatt "Bhavrang".

Dr. Rajeshwarji was recently recognized for his contributions to the field of Indian music by the Government of India by awarding him one of the highest professional achievement awards Padmashri. Acharyaji after completing his doctorate in music has taught many curious souls the intricacies of Indian forms of musical philosophy. In 1975 he revived the dying art of Dhrupad singing back then. Since then almost half a century has passed and through his untiring efforts in encouraging this ancient tradition, he has supported over 600 Dhrupad musicians and Mridang vadaks in the auspicious International Dhrupad Festival at Tulsi Ghat.

He has received numerous awards in his long life dedicated to Sadhana of Indian music. He not only is a great Dhrupad and Khayal singer himself but a world-renowned artist of Jal Tarang which is a rare form of art to produce music from water-filled bowls. He presided for over 24 years the music Department at DeenDayal Gorakhpur World University in Uttar Pradesh in India. He has conducted many research and scientific experiments related to music, which this author has had the pleasure of discussing with him. Through this ongoing dialogue of learning Pandit Rajeshwarji has showered his grace of knowledge on the author for which he is forever grateful. Acharyaji has always inspired this Author with his insights. One such deep perspective taught by Acharyaji will forever remain in Authors' heart;

"Music is Applied Philosophy"

This present effort to present in English the spiritual and philosophical dimensions of Swara, as well as its sensibilities, is a direct result of such inspiration and experimentation by the Author.

Sw. Shri Mannuji Mridangacharya

"Mannuji" was a Krishna devoted soul who unfailingly played Mridang during Kirtan Seva in Gopal Mandir in Benaras for Lord Krishna daily. His "Taal" was offered in the service of God. He was also the head of the Taal section teaching at the famous BHU (Benaras Hindu University). He learned Tabla and Mridang from Pandit Bholanath Pathak of Benaras. He was a very learned and selfless soul who made the rare effort to document his whole life's Taal lessons in a four-volume series called Taal Deepika in the year 1934 almost 85 years ago.

The present author was fortunate to have revived few rare remaining copies of these books from his relatives' house while they were literally rotting under a mattress. It is God's blessed Leela that this author has been able to translate a portion of Shri Mannuji's effort into English in his volume 'Science of Rhythm' and is dedicating it to the same source of all Mannuji's music, Lord Shri Krishna.

We do not have a photograph of Shri Mannuji else we would have surely published it here. The words that are flowing in this volume in English are a continuation of this great soul's devoted passion for Laya and his foresight of sharing it with others.

Sw. Mridangacharya Shri Ramshankar Pagaldas

(Pagal Baba), Ayodhya

Shri "Pagal Baba" as he was lovingly known was Pagal (Mad) in his single-minded devotion to the Taal Science and the art of playing Tabla and Mridang. He offered all of his musical abilities to Lord Shri Rama and resided in Ayodhya, where he taught many students as well. He was one of the rare Gurus who had the foresight like Shri Mannuji of Benaras, to document and synthesize his life's work in a book format for future enthusiasts like the present author. In 1965, He published his "Mridang Ank" with the complete theoretical and practical guidance for learning intricacies of Mridang from Sangeet Karyalaya, Hathras. His effort and his humility in writing the book and selflessly sharing all of his knowledge four decades ago is a testimonial to his

71

unflinching service to Taal Shashtra. Pagalbaba has been the author's first Mridang Guru in virtual presence through his "Mridang Ank". His soul is ever playing in front of Lord Shri Rama.

In one particular instance when discussing Taals, Pagal baba wrote in one of his notes that *"Taal Science is so deep that one Life might not be enough to master even a single Taal, let alone mastering All Taals and becoming an expert in all of the Taal Science."* This quote from him and its humility touch our hearts deeply. This author has not met Pagal Baba but our souls are indeed connected.

The flickering light of hope of the author's Mridang progress was kept alive by Shri Pagal Baba's selfless efforts in writing "Mridang Ank" many decades ago for precisely this type of inspiration for Laya seekers.

NL Shri Devakinandan Goswamyji
Indore

NL Shri Devakinandanji Goswamy also represents the lineage of Mahaprabhu Shri Vallabhacharya. This author met Shri Devakinandanji, by chance at Shri Nathdwara (Rajasthan) in 2006. In the temple of Shri Vitthalnathji (a form of Shri Krishna), where he was visiting for darshan, a Kirtan Pada was being performed depicting the mood of Raas and Taal was "Charchari". Shri Devakinandanji was performing Aarti of the Lord and the author was amongst the darshanarthis doing the darshan. Their eyes met and after the darshan, Shri Devakinandanji expressed the desire to meet this author and he blessed him with his two CDs, one of Dhamar Taal and one of Chautaal. His words and blessings will be with the author forever. He had instructed in that meeting; *"Counting of the Laya on one's fingertips will alleviate all stresses of the mind and take you to a place of eternal peace."*

Shri Devakinandanji was the one of the foremost Pakhawaj vadak of India while he performed on stage publicly and in privacy for his own Lord Shri Krishna. He performed Chautaal in the International Dhrupad Festival at Tulsi Ghat in 2010. While Author and Shri Devakinandanji met only once, he left his Tulsi Ghat

Chautaal recording as a milestone and guidepost for the Author. He learned Mridang from Nana Panse Gharana parampara and Shri Pannalal Pawar was one of his teachers.

'Rasik Pritam' Shri Hariray Mahaprabhu

Shri Hariray Mahaprabhu was the 5th generation Grandson of Mahaprabhu Shri Vallabhacharya (founder of Shuddhadvaita Krishna Bhakti Devotion path). He was a prolific musicologist and one of the most devoted authors composing hundreds of literary works and amazingly beautiful love-filled Poetries of Lord Krishna in Brij Bhasha, Sanskrit, Punjabi, Gujarati, and Hindi. He wrote most of his poetic works under the pen name of "Rasik Pritam" - (the great admirer of all the Rasa of Lord Shri Krishna.)

Shri Harirayji was intensely emotional in his approach towards Krishna. He was known for his humility and his profound devotion to his founding Guru (his ancestor and great Grandfather) Mahaprabhu Shri Vallabhacharya. All his life and efforts were dedicated to guiding himself and other souls towards the infinite

Rasa in the Lotus feet of Lord Krishna. He had a long and most peacefully revered life that graced the earth for 125 years. Having read most of his philosophical works as well as his poetry, the present Author is indebted forever to Shri Harirayji for inspiring and guiding his Naad Yatra.

Science of Melody

With blessings of Lord Shri Krishna, I hereby bow to
him to allow me to introspect, experience and explain
the inner sensibilities of the Science of Indian musical
Melody called "Swar Shashtra" for those souls who
are curious to learn and are hungry for creating and
partaking in the joy of the infinite musical
NaadBrahma. - Author

Science Art Philosophy Religion

The universal ideals of any music of any culture revolve
around two core fundamentals of Rhythm and Melody.
Rhythm's main mode of execution in the Indian system
is called "Taal" and Melody's main means of execution
in the Indian system is called "Swara". In the previous
volume published now in its second edition, we covered
the "Science of Rhythm" (Taal Shashtra). In the Indian
system of music, the Laya forms the "base" foundation
of all other forms of music and hence it was essential to
discuss the Science of "Laya" and Taal Shashtra in our
first volume. It is on this foundation of Laya that all

other forms of performing arts are constructed, be it Vocal music, Instrumental music or Dance. Of all of the three of these performing arts, it is said that Singing is the foundation for understanding the other two i.e. Instrumental and Dance forms. As such, it is now Lord's wish that we offer at his loving feet the "Science of Melody" in this second volume. Prior to understanding the details of the "Science of Melody" (Swar Shashtra), the author encourages the reader to glance through the section titled "Science, Arts, Philosophy and Religion" in volume 1 of this series. Revisiting an exact understanding of these four core areas of our lives is highly beneficial for a student of Indian Swar Shashtra.

It is, therefore, an ideal place for us to start by understanding what purpose of this "Science" here in the context of music means. For a soul starting their musical journey with Melody instead of Rhythm, we have decided to summarize this chapter as a starting point for "Swar Shashtra". For those who have read this introduction in the earlier volume, a refresher on the four elements would be beneficial as well.

The definition of all of the four major pursuits of life described above (Science, Art, Philosophy, and Religion) deal with mainly three aspects

1. First is the inner (physically unmanifested) universal operating laws of all creation around us,

2. Second is the outer (physically manifested) universe and creation all around us, and

3. The third is the "Creator" of these inner and outer natural laws as well as the physically manifested reality.

Science is the pursuit and thought process concerned with "Observing" and trying to "Quantify" the inner workings of Natural Laws. Science objectively tries and wishes as much as possible to "mold" these observations into our rational minds and define the "Physically manifested Natural World" around us in the confines of the "rational logic" of the mind. There is no place for "Unquantifiable" subjectivity and feelings in Science. The purpose of scientific pursuit is to identify in the rational confines of our mind the identity of who the **"creator"** of our physically manifested world is and the reasoning logic behind this creator and his creations. *Science deals with facts of our Life that can only be physically observed measured and repeated.*

Art is the pursuit of our human existence to "Experience" the emotions and feelings of the inner laws of nature. Art tries to connect through those emotions with the creator of these natural inner truths. Art is comfortable in dealing with the "Unquantifiable" sensibilities of our human existence. Things that are not confined in rational thought of our minds are acceptable in Art. Purpose of Art is to experience the Joy experienced in the inner unmanifested natural laws and personify this Joy in the Physically manifested world around us all through a "process" that is amenable to our Hearts. The objective of Art is ultimately to identify and connect with the **"creator"** of this physically manifested world and appreciate his creations through a process likable by our Hearts and emotions. *Art deals with Sensibilities of Life.*

Philosophy is the internal analysis of our human race that is concerned with observing the Outer physically manifested realities and then making efforts to understand the inner laws of the operations of this manifested universe. Philosophy deals with the faculties of rational thought in our minds and the pursuit of producing an **"observer"** philosopher who is aware and has mastered the knowledge of the inner natural laws of all things in the Universe.

Religion is concerned with "feeling" and connecting with the inner laws of nature through a process that is amenable to our Hearts "faith". The objective of religion in this sense is to produce an **"observer"** who can appreciate and partake in the infinite Joy existing in the inner natural laws of life. The purpose of religion is to know and experience the ultimate bliss and ultimately unite with this Bliss.

As we will explore in the book, music is one element of our human existence that is Universal in its nature. *Music has the power to encompass all the four dimensions of human existence as discussed above (Science, Art, Philosophy, and Religion).* Music is a means of approaching the unmanifested, manifested and the creator itself through its practice crossing all of the above four perspectives. We are striving in this volume to explore some of the Science (quantifiable logical aspects) and Artistic sensibilities (emotional) underpinnings of the Melodic "Swara" components of the Indian Classical musical system. As we analyze the Science of Indian Swar Shastra Author's frame of reference is certainly steeped in Philosophical and Hindu Vedantin spiritual spirit.

Swar Shashtra Philosophical Evolution

Any knowledge in the Universe is not created. *It is made available to the seeker who searches for it.*

According to Shri Shyam Manohar Goswamyji, "Shashtra" in the Indian philosophical system are the bodies of knowledge and processes containing various aspects of human existence and our entire universal being. There are many Shastra's and musical 'Swara Shashtra' is but one of these bodies of knowledge. The word Shashtra in Sanskrit is also related to weapons (Shashtra). Just as a weapon is used to attack in some manner to remove some negative (threat) aspect surrounding the weapon holder. Similarly, philosophical Shashtra are the bodies of knowledge that attack our ignorance and reveal the pre-existing truths to us.

In this context per Shri Shyam Manoharji, *Shashtra can never be taught or learned they essentially only can "Self – Reveal" themselves at the appropriate times when the seekers merit is ready for that inner knowledge.* What the Guru and Shishya are doing until then is preparing the ground to get these seeds of knowledge to bloom on their own, but the seeds of all knowledge already pre-exist amongst us and in the Universe.

All forms of musical systems in India, be it Vocal, Instrumental (non-rhythmic) or Rhythmic, are derived from the root of "Naad Brahman". In "Science of Rhythm", we have covered these key ideas to explore the philosophical underpinnings of the Indian musical Science. We have at length covered the precise and detailed definition of Nada Brahman and its Two Fold nature in terms of Aahat and Anahat Naad. We then described the concept of "Joy" in music and understood it to have its basis in "Resonance" within our modern scientific parlance.

We understood in great detail the "Psychosomatic" process of creation of Indian music. We also covered the spectrum of Naad in terms of what produces Joy in the performer's soul versus the aspects that produce Joy in the audience. We also analyzed music and its connections to Mindful Meditation as well as some observations from Neuro-Science about music. Lastly, we also reverse-engineered the process of music creation from the point of view of a musicians' execution of musical processes. All these ideas will form tremendously useful understanding for the present reader of this volume of "Science of Melody" as they are equally applicable to Melody as well as Rhythm. For the interested soul, we suggest those sections of Naad Yoga Volume 1 "Science of Rhythm" as good reference points to study as well.

Now in this present Naad Yoga Volume 2 "Science of Melody" we continue expanding upon this Scientific, Philosophical and Spiritual genesis in this context of Swar Shashtra. Prior to discussing the Swar Shashtra, we will now in this volume explore the complete Physical and Yogic process of the Human body and its components leveraged in the creation of music. One should not be surprised at this order of subjects since to create any audible music, one essentially needs some instrument and there is no better instrument in the Universe than the human form itself that we use to sing and experience music with. As such, this becomes one of key components of understanding the "Science of Melody". In order to produce Rhythm often we need to rely on external instruments and we need to understand those, but in terms of core essence of Melody and singing, the processes are internal to our body and those must be understood properly for any "Sadhak" of Indian music. Our human physical form is the first and foremost instrument of music, which must be understood in its precise design as it, pertains to the creation of Naad.

(vii) गीत-प्रशंसा

सामवेदादिदं गीतं संजग्राह पितामहः ॥२५॥

गीतेन प्रीयते देवः सर्वज्ञः पार्वतीपतिः ।
गोपांपतिरनन्तोऽपि बंशध्वनिवशं गतः ॥२६॥

सामगीतिरतो ब्रह्मा वीणाऽऽसक्ता सरस्वती ।
किमन्ये यक्षगन्धर्वदेवदानवमानवाः ॥२७॥

अज्ञातविषयास्वादो बालः पर्यङ्कुकागतः ।
इदमीतामृतं पीत्वा हर्षोत्कर्ष प्रपद्यते ॥२८॥

वनेचरस्तृणाहारश्चित्रं मृगशिशुः पशुः ।
लुब्धो लुब्धकसङ्गीते गीते यच्छति जीवितम् ॥२९॥

तस्य गीतस्य माहाऽऽरम्यं के प्रशंसितुमीशते ।
धर्मार्थकाममोक्षाणामिदमेवैकसाधनम् ॥३०॥

Importance and significance of the "Gitam" (Vocal music) i.e. melodic aspects of music. Vocal music origins are derived from one of the four Vedas, the "Samaveda" and it is none other than Brahma who collected these teachings of music from Samaveda through the "Samgiti". Omnipresent Parvati's husband Lord Shiva is praised in these Vocal songs. Lord Krishna is infinite and he is the Purushottam, yet for sake of his Loving Gopika, he becomes confined. Krishna's confinement of Love is through the means of many of his Leela plays of which the primary musical manifestation is the sound of the simple Bamboo flute "Banshi". When Brahma is enamored by the hymns of Samaveda the Goddess of supreme knowledge, Sarasvati is enamored by the love for her Vina, the stringed mother of all melodic instruments. Thus even the Supreme creator (His form in Trinity manifestation of Brahma Vishnu and Mahesh as well as his Shakti manifestations in form of Goddess Sarasvati, Parvati and loving Gopika) is completely bound to the melodic vibrations of his eternal creation of many Universes and these praises are sung in Vedas, what then are we to say about nature, animals humans, demons, divine beings and all other forms of creation in its entirety. They are all enamored by "Melody". The importance of Melody can't be described enough when it is the only "truly enjoyable" means of achieving four primary objectives of human existence per the Indian theosophy, i.e. Dharma (the righteousness of life), Artha (material well-being of life), Kama (desires of life) and Moksha (liberation from life). (25-30)

We start first our exploration of the subject by defining "Melody" as described over thousands of years in ancient Indian Vedic texts. The origins are to be found in the Samaveda and the creator Brahma himself extracted this music for all of his creation. As such music is Universal in its language and perception of joy regardless of societal, cultural as well as political and geographic or any other forms of human imposed divisions. The hymns of this Samaveda spiritual in nature and are sung in seven tonal notes identifying with what we call now seven notes. Essentially there are four main tones that get repeated in two lower and upper formats in these hymns. The tones of Samavedic Hymns are named as:

1. Krusta
2. Prathama
3. Dvitiya
4. Tritiya
5. Chaturtha
6. Mandra
7. Atisvarya

These form the oldest known record of Melodic structures recorded in Indian cultural heritage.

"Resonance"– Philosophical and Scientific

Since time immemorial, the philosophers and sages across the world have pursued their quest for understanding this Brahman (known by many names and many forms in different time, place and regions). To seek it and to be one with it. This same concept of Oneness is the internal purpose of all music whether it is knowingly pursued by the musician after studying the roots of this philosophical connection with Brahman or unknowingly pursued through just the act of singing or playing for Joy.

The fundamental scientific principle of "Resonance" is at work here. We know that when there are two different frequencies occurring close by to each other, the closer they come in their physical vibrating range, the closer they come to a point of "Unified Vibration" also known as the point of "Resonance". This is the point when the frequency gets amplified and the two separate frequencies become 'ONE' producing a single unified frequency.

When a musician or listener says that they feel they connected with the Spirit or God or The Supreme Creator, what they are essentially saying is that their "Aahat Naad" is finding a moment of "Resonance" with "Anahat Naad", the physically manifested and the all-pervading unmanifested subtle spiritual Naad Brahma unite and that is the climax of music. A commoner feels that music has this power but the common person does not know the science and philosophy of this Naad Brahma and its resonance. He/she can feel it when singing or listening and that is the Joy and Beauty of music and its Universality.

This **_Resonance_** between Aahat Naad (music that we can hear) and Anahat Naad (the music of Brahman's internal dialogue) is the ultimate objective of Oneness. A person singing, who with years and (often many a lifetime's) practice arrives at the perfect resonance with one of the seven musical pitches, he/she sometimes experiences roots of the hairs on skin tingled, or they have tears in their eyes. What is happening is that the singer is feeling this moment of oneness with the Brahman (the Creator).

Musicians also know this joy in arriving at the point of *"Sama"*. This Sama is a representation of this Resonance between Rhythmic and Non-Rhythmic cycles of music. The point of Union and Resonance and its Joy is infinite and impossible to describe in words, just as the infinite Brahman can't be described in our human words but can only be felt for the few who unflinchingly pursue this "Oneness" in the form of Naad Brahma.

Therefore, this is Naad in its elemental "Aahat" form that is heard by us. Naad in musical form that we hear is briefly of two formats:

1. Swara (connected with Vocal singing) – the Seven Notes comprising of their corresponding pitches used in Vocal music as well as Instrumental (non-rhythmic) music. Swara is used in singing or playing non-rhythmic instruments (Melody oriented). *Swara and Melody are subject of this present volume on Naad*

85

Yoga 'Science of Melody'. These notes (Swara), the science and art of producing those Melodious sounds and using our body as an instrument to produce those in singing are the subject of this book and this is known as the "Swara Shastra".

2. Laya (connected with Taal and its rhythmic beats) - The varied rhythmic sound produced by the rhythmic instruments. Laya is used in the creation and execution of the Taal. As present volume is focused on Swara (Melody), we will limit 'Laya' discussion to this brief definition. *Laya and Taal are discussed in-depth in a separate volume published earlier named 'Science of Rhythm'.*

Geometrical Science of Creativity

Three, Four, Seven and their Sensibilities

It has been a blessing for this Author to have learned musicologist and Padmashri Awardee Dr. Rajeshwar Acharya who represents the musical traditions of Late Pandit Omkarnath Thakur as one of his guides in learning and discussing many foundational elements of our Music. This scientific analysis of our musical structures aligned with numerical mathematics and geometry is one such blessing of knowledge that Dr. Rajeshwarji has bestowed upon the Author which is presented here.

Space and Melody Creation - For "Formation" of anything in nature what we need is basic elements of **"Space"**, "Execution" and "Storage". So to **"Create"** something we need Space. Space is essential for music. If we do not have space there is no music. The expansive nature of the sound element is essentially preconditioned on the existence of space. The format of sound in Melody is structured in some specific formations which we will explore in many details later but they are called; Sthayi, Antara, Sanchari and Aabhog. These create the boundaries of space in which the music is created and the sound element then expands (as per its execution) in that space.

Seven Notes

From a scientific point of view, the quantification of this space via geometry is essential for us to understand the creation of music. We know that a universally accepted format of music has seven notes. Be it Indian definitions of music, Western, or Middle Eastern or any other region. All music in human evolution is based on Seven Notes. The question arises as to why Seven? The answer lies in the fact that **two aspects that are needed to create music are Creativity and Space** as we reinforced earlier. These two key elements are represented in the laws of nature universally by two geometrical principles.

Four spatial corners – A square in Geometry is the basic foundational element of defining space. The four points of the corners of the square define the space that is being enclosed inside it. Therefore, in order for any space to be filled first, it must have boundaries defined geometrically and this is done through the four points of a square. This lays the FOUNDATION of the musical creation inside this space.

Triads ('Trik' in Sanskrit) of creativity – Now let us take three points and create a Triangle or a Triad. The triangle in Geometry is also considered as the strongest "form" or structure. In an aesthetic sense, this shape of the triangle also represents the true form of creativity. It is not only the strongest but most creative element of nature because it has the pointed upper point penetrating "space" which lays defined by our foundational elements of the four corners above. When we start from one corner of the triangle and depart, we have to leave the previous point and arrive at a fulcrum of the sharpness of the other point that acts to penetrate the space and create something new. (Great Pyramids of Egypt are visual examples of these aspects of strength and creativity). This interplay between 4 corners of 'Space' and 3 points of 'Creativity' geometrically keeps repeating in nature. The new elements of creativity always create new points of penetration. Remember where this 'creation' is occurring; it is inside the spatial confines defined by the square earlier. Therefore, to store this creativity we must have the four corners of the square to signify the spatial element where creativity exists and is stored and continues its evolution. *This 4 plus 3 natural numerical order of creativity is the foundation of the Seven Notes of melody.*

This is the sense of the natural laws of 7 notes the number 7 is composed of the eternally evolving Triangles (3) of creativity residing within the confines of the spatial dimensions defined by the Square (4). Thus, it is said that in every creation the element of 7 is needed. 7 notes of music, 7 colors and so on. This same principle was observed by ancient Indian as well as Western sages. What we now know as Golden Ratio in visual arts and mathematically as the Fibonacci sequence is also based on these same principles of Geometry. Take the same square we discussed as a spatial element and cut it diagonally and you will see that inside every space resides the yin and yang of two Triads of Creativity. This repeats ad infinitum and essentially relates to the concept in Vedic dictum;

"Anoraniyan Mahato Mahiyan"

The above profound evolutionary statement from Vedas is related to the laws of physics and science as well as manifest and unmanifest sound. It means that the Supreme Creator is all-permeating and is simultaneously the minutest point within the minutest form and the greatest vastness of space encompassing all the greater possible spaces. Some dimensions of this creator and his creation are manifest to us and some exist yet are unmanifest to us. Ano deriving from Sanskrit 'Anu' means individual soul (an animate or inanimate object), finest of fine particle, smallest of small atoms, minutest of minute form. Aniyat is unrestricted and Aniyaat in

Sanskrit means to become the minutest center amongst the smallest of small forms in the center of each soul and object. Mahat signifies the great and Mahaat indicates to become the greatest of great encompassing every great soul and object within its greatest boundary.

It is important to understand this root Vedic dictum before we discuss Musical creation. Let us consider the aspect of creation, space and form within natural laws further. The formation of any new element begins from the space in which form must be filled. However, where is this form coming from? It comes from each individual's unique relation to the supreme creator. This is akin to Einstein's theory of relativity. All creations are relative to the connections between the material creator of that element and their relation with the Supreme creator. In our discussion above take the square and triangle shapes and cut them diagonally or in halves and you will see they comprise of newer triads and spatial squares within.

Each one of these is connected to the main original starting form in some manner or the other but the experience of that form would be dependent on how one understands the relativity of that original relationship. This is then for the artist or musician to understand and keep experiencing this relativity himself or herself in every act of creation. The Gati (Speed) of creativity that we are exploring can increase the movement of time or decrease it as well. The entire experience in the space of time is relative to this understanding of the creative process by a musician.

Doyens of Indian music like Pandit Shri Kishan Maharaj and NL Mridangacharya Shri Devakinandan Maharaj understood this. In every performance of theirs, they played the Taal composition three times. When everything is in three wherever you go it will be Tihai and beginning of new creation. Because they knew the secret of logical and applied relativity of music and Philosophy.

The Triad "Trik" of Shadaj

Now let us explore the Swara and their formations. For many centuries now researchers in Indian Classical music have explained that originally, there were only three notes in the Vedas and everything evolved from there. It may be so but our approach is to go to a deeper understanding than this. First, in the context of Philosophy, let us understand that the Swar (notes) exist in the Universe on their own they do not need to be created and ALL the swaras exist with their relativity of relationships to others. To hear them distinctly in nature is also possible without identifying them individually but they all exist. It is the limitation of the human mind that without a name we cannot differentiate so first need in many different cultures was to name these natural Swaras.

On the Indian scale, the nomenclature is different and in Western, it is different but they mean the same thing and the base is the seven notes. Therefore, this naming aspect is a convenience for the human mind to better comprehend the nature of music. Secondly, once something is named, we still cannot visualize it without attaching some form to it and that is what the next immediate need is and hence the form of each Swara has been identified in the Indian system and further defined forms of groups of notes are the Ragas. All these terms will be discussed at length in this book.

Now in terms of Evolution, the Sound is composed of a trinity of Swarit, Uddat, and Anuddat. These can be also called the original "Trik" or the creative birth of sound in cosmology. All evolve from these Three. The Shadaj main note C on the western scale or Sa on Indian scale is called Shadaj. In Sanskrit, it is composed of two sounds both equaling a "Trik" sound of 1.5 measure each creating the main word 'Sh Ad Ja'. From this Shadaj all else is born. Let us explore how.

This melodic birth process utilizes the Swarit Uddat and Anuddat terms. Basically everything is relative to the place where you start and in sound, if Shadaj is our first Swarit note then we must go the next higher up one which is Uddat and as soon as we do that now we have two points which can have a middle which is slightly lower and that is Anuddat. So each note can thus have its own Trik of creation. In the concept of "Murchana" in Indian music, a student can sing Sa from any note with this idea of three.

So if Sa is Swarit, and Ga is Uddat then Re becomes Anuddat. Taking us to the first original scale in which all else can be contained but if we take it one octave higher than the same Sa Re and Ga can be experienced in Pa Dha and Ni a second Trik that is slightly higher than first Trik. So now, we have two Triads of 6 Swars and the moment we have two we have a middle and that becomes our Madhyam M note. In this way, the natural sound evolves and it expands using our Triad (Trik) of creativity. This system is an ideal system of management of Sound and its scales because it naturally evolves in the creative formation of Swarit, Uddat and Anuddat formations. These are the origins of Sounds in Samaveda and they start from the system of three. To say that seven notes do not exist in nature from our viewpoint is slightly misleading but the idea is to understand their genesis with the perspective of 'Triad' of creativity as we have just discussed

Science of Melody Reference

Ten Prana of Melody & *Ten Lakshana* of Jati

	Prana Name	Macro Grouping	Secondary Category	Practical musical/physical Sub-Category
1	Naad	Anahat (Unmanifested Conscious Sound) Aahat (Manifested musical Sound)	Aahat Navel (Ati Sukshma – Extremely Subtle) Aahat Heart (Sukshma – Subtle) Aahat Throat (Pusht – Loud) Aahat Cerebrum (Apusht – Faint) Aahat Mouth (Krutrim – Modified)	Aahat Heart (Lower Octave - Mandra) Pitch Intensity 1 Aahat Throat (Medium Octave – Madhyam) Pitch Intensity 2 Aahat Cerebrum (Higher Octave – Taar) Pitch Intensity 4
2	Shruti	22 Grouped as	4 Shruti Interval (4 x 3notes=12) (s m p) 3 Shruti Interval (3x 2notes =6) (r d) 2 Shruti Interval (2x2notes = 4) (g n)	Heart (Mandra Register) 22 Throat (Madhyam Register) 22 Cerebrum (Taar Register) 22
3	Swar (Shruti) 7 Pure Swara and 12 Vikrit Swara	Shadaj (4) Rishabh (3) Gandhar (2) Madhyam (4) Pancham (4) Dhaivat (3) Nishad (2)	Throat Root of Palate Lips Center of Cerebrum Teeth plus all of above Throat and Palate Throat and Lips	"Brahmagranthi Yogic Center of Body" Navel Heart Throat Root of Palate Cerebrum Sahastradhara Chakra
4	Gram 2 main	Shadaj (Human-Deshiya)	Brahma	Pancham is 4 Shruti

	Grama	Madhyam (Human-Deshiya)	Vishnu	Pancham is 3 Shruti & Dhaivat is 4 Shruti
		Gandhar (Divine-Margiya)	Mahesh	
5	**Murchana-Krama** 4 Fold Group 14 Pure	Shudh Kakali Antara Kakali Antara	14 Pure Murchana	7 Pure Shadaj Grama 7 Pure Madhyam Gram
6	**Taan** in Shadaj & Madhya-Grama	Shudh Taan Shadav Taan Audav Taan Kuta Taan	14 (7 +7) 49 (28 +21) 35 (21 + 14) 317,930	
7	**Sadharana** 4 fold	Kakali Sadharana Antara Gandhar Sadharana Shadaj Sadharana & Madhyam Sadharana		
8	**Varnalankara** 4 fold Varna	Sthayi Arohi Avrohi Sanchari	7 Sthayi 12 Arohi 12 Avrohi 26 Sanchari	Varna = Tonal Pattern Alankara = Embellishments
9	**Jati** 18 Jatis	*Shadaji* Shadaja Kaishiki Shadajotyava Shadaj Madhyam *Arshbbi*	7 Pure Jatis (named after their main Graha Swara in italics in left) 11 Modified	Melodic Type structures which give birth to what we know as Ragas defined by the corresponding **10 Lakshana.** 1. Graha – Initial note 2. Ansha – Fundamental note

		Dhaivati
		 Naishadi
		 Gandhari
		 Madhyama
		 Panchami
		 Gandharoditchyava
		 Raktgandhari
		 Kaishiki
		 Madhyomidtchyava
		 Karmaravi
		 Gandharpanchami
		 Andhri
		 Nandayanti
				a. Antarmarga
				 3. Tara – High pitch
				 4. Mandra – Low pitch
				 5. Nyasa – Final note
				 6. Apnyasa – Semifinal note
				      a. Samnyasa
				      b. Vinyasa
				 7. Bahutva – Profusion
				 8. Alpatva – Rareness
				 9. Shadav – Hexatonic
				 10. Audav - Pentatonic

10	**Giti** 3 Fold	Padashrita Talashrita Swarashrita	Magadhi ArdhaMagadhi Sambhavita Prithula	Kapala and Kambala Ancient Giti

96

Section 1 - Sense of Musical System

Having understood the Philosophical background of music, Taal and Swara, we now explore the precise inner psychological process as well as physiological process in producing the sound element of our melody. This process is described in much detail in the Indian Vedic scriptures and Upanishads, but we summarize it here to understand our subject of musical melody in its entirety. This process and its execution with spiritual goals is the starting point of Naad Yoga being introduced in this section.

Music - Science and Spirit of Naad Yoga

"Yoga" has been the gift to humankind coming from the deepest origins of Indian philosophy and the Hindu Vedas. The type of Yoga that interests the seeker of Melody here is "Naad Yoga". The term Yoga in general in Sanskrit means "to Yoke" or "to join". In Naad Yoga, one uses the format of Naad and its processes and science to unite with the Creator (Brahman) and his many aspects of realities leading to itself.

All Yogic practices have a prescribed format and structure within which one must execute these to achieve some desired results. **The Naad Yoga in its simplest form is practice of music with a Spiritual objective in mind. This musical practice could be any form of music (Vocal, Melody or Rhythm or even Dance) so**

long as it has the oneness with the *Brahman* as its objective. The practice essentially is a process of becoming aware of the universe around us gradually appreciating it with its many variations and ranges with Naad "Sound" as the entry point. Those universal infinite possibilities take manifestations in forms of Swara and Laya becoming "alive" in the experiential sense for the Naad Yoga practitioner. As in all Yoga practices that are focused inwards, the practice is not a social one with many people but rather is best performed in front of one's spiritual point of focus. By its very definition, the practice and experience of Naad Yoga occurs in the individual pursuit and is not a collective experience.

Naad and its Two Fold Manifestation

अथ द्वितीयं पिण्डोत्पत्तिप्रकरणम्

क. विषयावतरणम्

(i) नादमहिमा

गीतं नादात्मकं वाद्यं नादव्यक्त्या प्रशस्यते ।
तद्द्वयानुगतं नृत्तं नादाधीनमतस्त्रयम् ॥ १ ॥
नादेन व्यज्यते वर्णः पदं वर्णात्पदाद्वचः ।
वचसो व्यवहारोऽयं नादाधीनमतो जगत् ॥ २ ॥

Importance of Naad. The "Primordial Sound" that we define as Naad is the formative existential essence of Vocal music. Instrumental music becomes pleasant and enjoyable due to this Naad. Moreover, Nritya (Dance) is dependent on both Vocal; Instrumental forms of Melody, and as such, it cannot exist without Naad. Thus, all the forms of music depend on Naad for their existence. Naad is to music what Breath is to our Life. (1)

Naad sound manifests the "Varna" (the alphabetical letters), these then form words, and words constitute sentences of communication all dependent on Naad. The entire business of life is conducted through this communication in the form of languages and their tones. As such, the whole world and its phenomenon is in existence due to the Naad. (2)

We described earlier, the importance of Melody in the Indian system. Now we start by revisiting the Naad and its importance for Melodic music. We have already covered Naad earlier in previous volume on Rhythm to some degree. Now we describe the other characteristics that allow us to understand it even more deeply. Philosophically Naad is nothing but Brahman the Supreme Reality responsible for creation and these aspects are covered in Volume 1. Here we are now entering the connection of the unmanifest creator into its manifest reality of Naad and analyzing its material aspects as well.

Indian system of philosophy, considers five main elements "Panch Mahabhoota" of our existing world and these are corresponding to the five perceptible senses of our human body. (Panch Indriya).

1. Earth - corresponds to our sense of Smell
2. Water - corresponds to our sense of Taste
3. Fire - corresponds to our sense of Vision
4. Air - corresponds to our sense of Touch
5. Akash (Ether) - corresponds to our sense of Hearing

Of the above, the Akash has no precise English word that defines its unique Sanskrit meaning. Closest we can come is Ether and as such we can understand that the first four elements of the world are all encompassed in this all-pervasive and all-encompassing Akash (Ether). The unmanifest Creative element of the Universe manifests itself in many unique forms that are known by many different names. Forms are based on creators' wish and one of its unique forms of manifesting itself in Akash is through its Naadatmak Sound. Since Naad itself is considered to be Brahman the creator itself, it manifests in audible sound format through the Ether and we can perceive it through our sense of Hearing. This Naad as described in the verses above essentially forms the basis of our world and its existential realities.

(ii) द्विविधो नाद:, पिण्डोत्पत्तिप्रतिपादनौचित्यम्

आहतोऽनाहतइचेति द्विधा नादो निगद्यते ।
सोऽयं प्रकाशते पिण्डे तस्मात्पिण्डोऽभिधीयते ॥ ३ ॥

This Naad is of twofold variety. Unmanifested or unproduced and Manifest or produced. The manifest audible Naad is produced in the required circumstance and instrumentation of Human bodily form. As such to study Naad and its evolution one must study in detail first the Human form itself. Our human form is the first materially manifest instrument of Naad production. (3)

Now we build upon ideas of Twofold Naad we discussed earlier in Science of Rhythm. To summarize there are two types of Naad.

1. Anahat Naad which is an unmanifest form of Brahman and
2. Aahat Naad which is the audible manifest form of Naad Brahman.

In the book henceforth all our discussion would elude to this second variety of Naad which is the manifest format of Naad that we can hear as "Dhvani" sound. This is the one that interests us from the point of view of studying the "Science of Melody".

The Aahat Naad is manifested first as pure 'Dhvani' (sound). Then it gains more linguistic meaning when it associates itself with 'Varnas' the language alphabetical meanings. When coupled with these linguistic sounds that can be heard from a distance and differentiated in terms of their meanings through the tonality of pitches the Naad becomes a means of communication as well as melody.

Metaphysical Dimensions

We now further explore the metaphysical aspects of the evolution of this Naad. This was explained in Science of Rhythm from an evolutionary perspective, now we continue this dialogue and extend it to our Human form.

The questions that one might ask is; Why study these physical body connections in a book about music and Science of Melody? What does the human physiology and genesis of human embodiment all the way from conception in the womb have to do with music? The answer lies in understanding the fact that prior to studying music one must study the instrument of its production as well. Sharangdev who was the author of Sangeet Ratnakar which forms the basis of our analysis was unique in that he was not only a musician but was an accomplished physician and practitioner with deep knowledge of the human body and Ayurveda (ancient Indian system of medicinal knowledge)

This allows us to explore music and Melody from its entire spectrum of Philosophical, Physiological, Scientific and Artistic perspectives. The human body is part of the existing universal metaphysical matter and corresponding phenomenon. The human body is the greatest instrument that has ever been created by Providence. Our physical form; naturally and freely available, is the 'penultimate' means of Naad production. As such understanding aspects of Human embodiment and its connection with Creation and Naad are both essential prerequisites for the true seeker of understanding the full sensibilities of Melody and music.

To understand the evolution of Naad we go back to the source, the roots of the evolution of the Universe itself as we did earlier in Volume 1. We now continue that journey in further detail. Order of our Metaphysical evolutionary analysis is as follows;

1. Nature of Brahman the Creative Supreme

2. Nature of the Individual Being (Jiva/Soul)

3. Evolution of Subtle Body

4. Universal Cycle of Creative Destruction

5. Jiva/Soul and Brahman relation

6. Material world

7. Physical bodies of the material world

8. Naad related Embodiment of human form

(i) ब्रह्मस्वरूपम्

अस्ति ब्रह्म चिदानन्दं स्वयंज्योतिर्निरञ्जनम् ।
ईश्वरं लिङ्गमित्युक्तमद्वितीयमजं विभुः ॥ ४ ॥

निर्विकारं निराकारं सर्वेश्वरमनश्वरम् ।
सर्वशक्ति च सर्वज्ञम,

'Brahman' is the self-luminous essence of all creation.
It is the 'Is'-ness of existence. Characteristics of this
Brahman are infinite but few listed here are; (4)

1. Blissful

2. Awareness

3. Taintless

4. Supreme Being/Deity

5. All-encompassing

6. Penultimate Cause of all creation

7. Penultimate Cause of all destruction

8. Unborn

9. Non-Dual (Singular Reality)

10. Infinite

11. Unmodified

12. Formless yet with Many Forms

13. Imperishable

14. Omnipotent

15. Omniscient Timeless

16. Supreme Ruler

This Brahman in the "Saguna" form is with qualities that we can identify with. This form of Brahman that we call God or "Ishwara" is the one which is qualified with universal consciousness and is worshipped by humans through the ages in many different forms and formats depending on time, place, region, material circumstance, etc. This Brahman in the form of worshipped divinity is the one that controls the evolution of our material and spiritual universes. It is capable of doing, undoing and doing that which opposes the expected course of action. Brahman is the Supreme source of ALL manifestations.

(ii) जीवस्वरूपम्

, तदंशा   जीवसंज्ञकाः ॥ ५ ॥

अनाद्यविद्योपहिता   यथाऽग्नेर्विस्फुलिङ्गकाः ।
दार्वाद्युपाधिसंभिन्नास्ते   कर्मभिरनादिभिः ॥ ६ ॥

सुखदुःखप्रदैः   पुण्यपापरूपैर्नियन्त्रिताः ।
तत्तज्जातियुतं देहमायुर्भोगं च कर्मजम् ॥ ७ ॥

प्रतिजन्म प्रपद्यन्ते,

'Jiva' is equivalent to the word soul in English. This Jiva is nothing but an infinitesimal 'part' of the Supreme Whole 'Brahman' In this manner, Jiva and Brahman are connected in a manner analogous to the fire producing Sunrays (Jiva) connected to Sun (Brahman). Just as sparks of fire from a wood produced from the Sunrays are differentiated in circumstances of the wood, Jiva while all connected to Brahman are differentiated by the nature of their circumstances of individual merit and evolution.

Jivas are governed by their actions that are beginingless from infinite time. Actions both good and bad and corresponding to pleasure and pain, the Jiva thus attains birth in a physical manifestation associated with past actions of their accumulated 'unfructified' karmic cycles. (5-7)

Brahman is 'timeless'. It is infinite. Time is a relative concept and its value is only in the relative existence of the matter which is being discussed. In absolute terms, this time has no value, as it is Timeless. Actions of Brahman are timeless and so are the Jiva produced as partial manifestations of Brahman itself. However, the actions of Jiva are relative to their cycles of evolution and dissolution based on the differentiation of their individual conditioning.

Thus even though all individual beings are evolved from Brahman, each is unique in its conditioning and its uniqueness of past accumulated "Karma" actions. This then forms the basis of the Law of Karma that guides the Jiva into the cycles of Birth and Death. Karma are incomplete partial actions across time-space that necessitate the relativity of Time for its completion. This is similar to a seed that is sown which then grows into a tree. But the seed needs time to grow into a tree and as such planting of the seed is also part of the ultimate process of fruit being born on that tree.

The key summary is that the Jiva and Brahaman are related however the Jiva is bound by its actions (Karma) which are beginingless since Brahman is beginingless. However, the uniqueness of the actions of each Jiva is per the characteristics as wished by the Brahman. Good and evil actions of the Jiva are both products of this beginingless Brahman's creation, but they are not endless. The Jiva can end the cycle of these actions through seeking and attaining the right knowledge of its relation with Brahman and through self-awareness thereby self-correcting many of past non-optimal actions. This essentially is the path of liberation. Thus no Jiva is damned eternally and every soul has a Hope for redeeming their true relationship to Brahman that they might have over a course of time cycle forgotten.

(iii) सूक्ष्मशरीरम्

, तेषामस्त्यपरं पुनः ।

सूक्ष्मं लिङ्गशरीरं तदामोक्षादक्षयं मतम् ॥ ८ ॥

सूक्ष्मभूतेन्द्रियप्राणाऽवस्थाऽऽत्मकमिदं विदुः ।

Having discussed the Brahman and the Jiva which essentially form the first singular link of evolution. We now start going further down the chain of evolution. Let us explore the next step of metaphysical evolution. This is the 'Subtle' Body. Jivas are comprised of two aspects of their bodies the "Subtle" and the "Gross" (Physical body which is to be discussed at length in the next section)

The 'Jiva' is composed of a 'Linga Sharira' that in English translates into 'Subtle' body. This is indestructible until the Jiva is emancipated and merges back into Brahman. The 'Subtle" body is composed of the following; (8)

1. Five Great elements - (Panch Mahabhoota) (Earth, Air, Water, Fire and Akash/Ether)

2. Five Senses - (Panch Indriya) (Smell,

3. Vital breaths in their subtle form

Let us remember that all Jivas are essentially partial manifestations of Brahman which is eternal so how do they pass through the evolution and dissolution cycles of birth and death. This is understood properly once we examine that the Jiva is comprised of not just "Gross" physical body which definitely must decay and perish over time, but it also is composed of a much finer and infinitesimally subtle body made of subtle matter consisting of the Five great elements.

This subtle form is imperceptible by our senses except for the Yogis and sages who through years of rigorous study and experiments of Yoga can perceive their subtle bodies. This subtle human body is the one which is eternal just as is Brahman until the point of Liberation when this body ceases to exist and merges with Brahman itself as Pure Consciousness. But for all practical purposes prior to Liberation of that Jiva, the subtle body never dies it just passes from one physical body and form to another taking with it the accumulated actions and their continuing conditions from one birth to another. This is the metaphysical process of the Indian cycle of rebirth.

(iv) सृष्टिसंहार-प्रवाहः

जीवानामुपभोगाय   जगदेतत्सृजत्यजः ।। ९ ।।
स आत्मा परमात्मा च विश्रान्त्यै संहरत्यथ ।
तदेतत्सृष्टिसंहारं   प्रवाहानादि संमतम् ।।१०।।

The two states of "Subtle" body and "Gross" physical body necessitate us to study the purpose of creation 'birth' and destruction, 'death'. This is explained here in the Cycle of the Universal forces of creation and destruction. Here we study the purpose of creation itself. The process has been described in the Science of Rhythm.

The 'unborn' Brahman creates all that is existing for HIS enjoyment and his play 'Leela'; Individual souls 'Jivas' being part of Brahman itself also partake in this enjoyment as the world is created for them and with them. Once created, the 'Atman' (Universal Self) then withdraws his creation back unto himself for sake of Rest from the act of play. Thus then continues the eternal cycle of creation and destruction; activity and rest, birth and death. (9-10)

Thus far we discussed what is the nature of the Supreme Reality (Brahman), How is this supreme Reality related to us individual beings. In this verse, we explored the relation of the world itself with the Supreme reality. In one word we can summarize this relationship as that of 'Leela Bhava', natural act of joyful play. This play is constant in its cycle but stages of play are cyclical and transitory. This Leela phenomenon is continuous in that it comes into initiation, continues for a while and then returns to its source. The 'Source' is timeless and beyond causal logic i.e. causeless and eternal.

Creation has no motive, it is spontaneous out of the wish of the Supreme Brahman. The word 'Atma' here must be noted is equivalent to Brahman, the Supreme Reality. However when we use it in the context of prefix 'Param' + 'Atma' = 'Paramatma' we also refer to the reality as the Universal God or 'super self' more as a perceived divinity. When we refer to the same Atma with prefix

'Jiva' + 'Atma' = 'Jivatma' we then refer to the individual embodied soul or the individual self.

Such is the interplay between the Supreme Lord's manifest and unmanifest phenomenon - an eternal play of the Lord – Leela.

(v) जीवब्रह्मणोः सम्बन्धः

ते जीवा नात्मनो भिन्ना भिन्नं वा नात्मनो जगत् ।
शक्त्या सृजन्नभिन्नोऽसौ सुवर्णं कुण्डलादिव ॥११॥

सृजत्यविद्ययेत्यन्ये यथा रज्जुर्भुजङ्गमम् ।

We have analyzed the nature of Supreme reality Brahman, the Jiva and the cycles of creation and destruction. We now observe that all these various facets are uniform in their singularity.

The individual souls 'Jiva' is essentially part of this Material World, as they constitute the world. Both Jivas and the world itself are nothing but part of the Supreme Brahman itself and thus they are all related in oneness to the Supreme Creator. In other words, they are all related to the singularity of creation and are 'Nondual'

Just as Gold is non-different from its forms of jewels, in the same way, Jivas and the World are non-different from their Supreme Creator 'Brahman'. (11)

This is the 'Non-Dualistic' "Advait" Vedantin approach where all is ONE with the Creator and there are no false perceptions. Every facet of creation is real and one with the creator. Here we leverage the very lucid description from Shri Shringy's translation. "The relationship of the Supreme Creator, the created individual beings and the World is that of 'identity with reference to substance' and 'diversity with respect to forms'." Thus all music is HIS creation and is essentially ONE with Creator himself in this model of Non-Duality. Just as Gold Ornaments are Gold in substance but carry different names and identities of the diverse jewel forms.

(vi) भौतिकसृष्टिः

आत्मनः पूर्वमाकाशस्ततो वायुस्ततोऽनलः ॥१२॥

अनलाज्जलमेतस्मात्पृथिवी समजायत ।
महाभूतान्यमुन्येषा विराजो ब्रह्मणस्तनुः ॥१३॥

Treatment of Svara

ब्रह्म ब्रह्माणमसृजत्तस्मै वेदान्प्रदाय च ।
भौतिकं वेदशब्देभ्यः सर्जयामास तेन तत् ॥१४॥

तदाज्ञयाऽसृजद् ब्रह्मा मनसेव प्रजापतीन् ।
तेभ्यस्तु रेतसो सृष्टिः शरीराणां निरूप्यते ॥१५॥

We have established many different aspects of the manifested world before us in stages earlier; The Brahman Supreme Reality and its relation to the soul; the manifest cycle of creation and destruction and the singularity of all of these relationships to the Supreme Being. Now we focus on the specific process and order of 'Creation' of this Physically manifest natural world.

The seeds of all creation lie first in the five great elements that emanate from the 'ParamAtman Supreme Creator. Five great elements 'Panch Mahabhuta' are said to form the body of the Brahman 'Viraj' itself and emanate in the following order;

     I.   Ether (Akasha)

     II.   Air (Vayu)

     III.   Fire (Agni or Anal)

     IV.   Water (Jal)

     V.   Earth (Bhoomi or Prithvi)

Ether is closest to 'Akasha' of Sanskrit language and hence used here even if not completely representing the full meaning of Akasha. This also is the first element from which all others evolve and is an all-encompassing element.

The Supreme Creator 'Brahman' manifested itself through his body into the form of Lord Brahma who is entrusted with the task of creating the Universe. (Lord Brahma along with Lord Vishnu and Lord Shiva form the Trinity of Deities representing the Supreme and Singular 'non-dual' Brahman.)

The Supreme 'Brahman' then hands over the 'Scriptures' the Vedas to Lord Brahma. Through the verses of Vedas as seeds of creation, Lord Brahma creates the physically manifested world. From his mind and its power of imagination and will, Lord Brahma creates many 'Prajapati' from whom seminal creation of physically manifest bodies is initiated through physical means of reproduction. (12-13)

Lord Brahma is also called by name 'Viraj' in Vedas meaning simply the one who is Luminous. In the first instance of this creation, Brahman itself creates the five great elements and then 'enters' into those elements and from the body of 'Viraj' who is the manifested form of Lord Brahma. This due to principle of 'Advaita' singularity there is no difference at all between creator and creation.

Each of the five great elements is responsible and corresponds to interact with our individual 'Jiva's five senses 'Indriya'. These five elements are responsible for combing in many permutations and combinations to create our physically manifest material world made up of their matter. *Akasha is the medium and substratum through which Sound which concerns us most importantly in Science of music is propagated and perceived through our ears.*

Vayu is the element through which perception of touch occurs. Agni is the medium through which perception of color and vision takes place. Jala is the element responsible for the perception of taste. And finally, Prithvi is the element that is responsible for our perception of smell through the nose.

Brahman itself is formless however when it forms the five great elements and wishes to create the Universal manifestation, it takes the form of Lord Brahma who has a form with four faces signifying the four Vedas that Brahman gave to him. The actual creation of the physical world then occurs through Lord Brahma using the seeds of words from the Vedas. The Vedas themselves are considered to be unborn and eternal and are called 'Apaurusheya' in Sanskrit meaning beyond human authorship.

In the "Advait" philosophy of Singularity creation and dissolution both are unitary and corresponding actions. As such the seed words of Vedas 'Create' the world through association giving rise to a 'form' in the Universal Mind. This Universal mind, in turn, manifests this material world with forms of those ideas of Vedic words as seeds. What is the reason one must ask at this juncture for this creation? Simple as we have discussed earlier, the reason is the 'Play' or 'Leela' bhava of the Supreme Brahman.

Since the audible sound element of this universe is one form of Naad, it propagates through the first of the five great elements Akasha in this physically manifested world as described above. This Naad thus is perceived, received and heard through the substratum of Akasha into our ears and into our human physically manifest body form.

Naad - B flat 57 octaves below middle C

Musical Knowledge is not created in the Universe, it pre-exists. In other words, Sound Element is never produced. "Naad" Sound Element is essentially made available by the Providence in creation itself as a preexisting requirement of creation. However, after much guidance from a learned Guru and committed practice and humble labor, musical Shashtra is self-revealed at some point of time in the true seeker's quest and the music is essentially made available from that point on in its truest form for them.

Sound Element (Naad) and Space Element (Akash)

As we have seen, Naad is twofold one that is manifest and can be heard (Aahat Naad) and one that is unmanifest and unheard by human ears (Anahat Naad). The Indian sages knew of this since this is described in the evolutionary theories of the universe within the Vedas; however, a question might arise in an inquisitive mind as to what is the proof of two-fold Naad. Manifest Naad can be heard so we know it exists. Our focus here is on exploring the 'unmanifest' quality of the Anahat Naad.

There are several types of proofs in the evolution of knowledge of a subject per the Indian system of philosophy.

1. **Proof as expounded by the Vedas** – We have already covered this in the statement above depicting the Twofold nature of Naad. The Vedas identify this as such and this is the profound proof in Indian philosophy.

2.   **Experience-based proof** – Many concepts of science and spirit can be realized only through personal experience. Anahat Naad is one such aspect. Sages of yore through meditation and self-discovery process over millennia across many different spiritual disciplines have reached the same conclusion on the primordial nature of Anahat Naad. The Indian rishis and sages through meditation can experience, 'hear', and be one with this Anahat Naad and this for them is not a musical exercise but rather an act of connecting with the Supreme Creator himself.

Another great example of this model of experiencing the Anahat Naad is the chant of the Buddhist Monks in the Himalayas and the Tibetan plateau. The Author having witnessed this in their monasteries can personally attest to the deep practice of daily chants of the Buddhist scriptures that occur at such low frequencies that common people are unable to speak or sing at such frequencies. Their objective being to sing those frequencies and connect with the 'infinity' that is the primordial nirvana for their Buddhist spiritual path.

One must note that the lower the audible frequency. The closer we get to the unmanifest quality of sound and hence reach the Anahat stage. Sound in lower frequencies expands and permeates the whole environment and this is the objective of the Buddhist chants often that are practiced at the lowest of lower human octaves. These things can only be explained by practical experience and observation. Similar lower frequency usage is observed even in the western concept of Gregorian Chants where the "Naad" produced is of much lower octave than normal usage.

3.   **Guru Shishya tradition-based proof-** Another form of understating the proof of this natural

phenomenon is through the systematic and traditional Teacher and Student (Guru-Shishya) traditions in any culture where the words and experience of the Guru guides the Student and the student uses the teachings of such Gurus as described above as proof of Anahat Naad.

4.   **Proof as observed through Modern Science** – It is quite understandable that some modern souls would only accept the modern scientific proof. Concerning this fact, we need not look further now. Modern science has been able to accidentally arrive at findings that validate our concept of what we know in Indian philosophy as Anahat Naad. Let us explore this as well in terms of auditory sense, putting aside the spiritual and philosophical aspects that science would not consider as compatible elements.

In the year 2003, a team of international scientists and astronomers led by Dr. Andrew Fabian at the Institute of Astronomy in Cambridge, England, used Chandra X-Ray observatory of NASA and published their findings in a paper in 'Monthly Notices of Royal Astronomical Society'. Chandra X-ray Observatory is an orbiting X-ray telescope that sees the Universe in X-ray light just as the Hubble Space Telescope sees it in visible light. Their objective was to study Black Holes and their evolutionary life cycle. While studying Black Holes, this team of scientists documented and observed in their paper, the existence of what can only be termed as primordial sound in the Universe that was detected by their team.

Their observations noted that they were able to detect the sound of a black hole singing. Moreover, a sound that was detected, which the Black Hole has been probably singing for more than two billion years, is B flat - **a B flat 57 octaves lower than middle C.**

Black holes, as explained by Einstein's general theory of relativity, are objects in space so dense that neither light nor anything else, including sound, can escape them. But long before any sort of material disappeared into a black hole, theorists have surmised, it would be accelerated to near-light speeds by the hole's gravitational field and heated to millions of degrees as it swirled in a dense doughnut around the gates of doom, sparking X-rays and shock waves and squeezing jets of energy and particles across space.

This Black Hole 'B Flat' note is as close to Anahat Naad that modern science has been able to get. The experiments were focused on studying a supermassive Black Hole in Perseus cluster of galaxies, 250 million light-years distant. The "notes" appear as pressure waves roiling and spreading because of outbursts from the black hole through a hot thin gas that fills the Perseus cluster. These sound wave notes are 30,000 light-years across and have a period of oscillation of 10 million years. **By comparison, the deepest, lowest notes that humans can hear have a period of about one-twentieth of a second.** The Black hole in the Perseus cluster is essentially playing the **'lowest observable auditory note in the Universe'.**

Astronomers suspected that black holes in the central galaxies of clusters might be keeping the cluster gas hot, but the astronomers did not know-how. A particularly massive black hole is believed to lurk in a galaxy known as NGC 1275, which lies at the center of the Perseus cluster.

As the brightest X-ray cluster in the sky, radiating 1,000 times as much energy in X-rays as visible light, Perseus is a logical laboratory for investigations. Two jets of radio energy shooting out of the galaxy's nucleus have blown two bubbles in the gas in the center of the cluster. In an X-ray image from the Chandra satellite released in the year 2000, these bubbles looked like the eyeholes of a giant eerie orange skull.

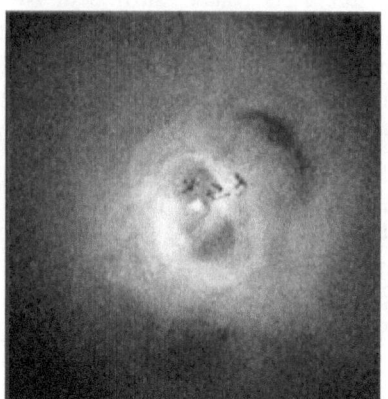

Three color Chandra Observation of Perseus Cluster Courtesy NASA, USA

A long-exposure Chandra image of the Perseus cluster, showed waves moving outward like ripples on a pond from the central bubbles.

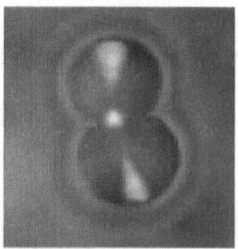

Still image from animation of sound waves generated in Perseus cluster: Courtesy NASA, USA

Researchers compared the process of gas movements to a child's blowing bubbles in a glass of water through a straw. In this case, the jets are the straw. The bubbles pushing against the enormous pressure of the gas surrounding them create sound waves moving out through the cluster's gas, pumping energy into it and heating it.

The energies responsible for producing this sound are phenomenal. It takes the energy of 100 million supernova explosions to blow a central bubble in the Perseus cluster. If the black hole blows such bubbles continuously and it is this energy that has been keeping Perseus hot, then the black hole in Perseus must have been playing a steady B flat for a long time per the research. **'This is the symphony of Creator and his Creation.'** This scientific discovery is just one famous scratch of the surface. There are yet unsolved mysteries related to Black Holes, Singularity and the fields of quantum physics that in our future generations will shed much more light on these subjects of evolution and Anahat Naad in the future.

These are a few of the approaches to understanding 'The' Anahat Naad. Many things in life **'exist'** in such a manner, that our senses are unable to comprehend them fully. Things exist yet we cannot experience them through our physical sensory dimensions. Above scientific discovery of primordial 'B Flat notes' emanating from a Black Hole prove to us that the 'unmanifest' nature of Anahat Naad is a fact-based on 'experiential' learnings of sages and scientists of our past. Indian musical science identifies Anahat Naad as the sound that is 'unheard' by human ears, yet it is the main support of the Universal cycle of creation and evolution. Through this fascinating observation of modern science, it has reached one step closer to the idea of our 'Anahat Naad'.

Even though this is the lowest frequency of sound that our modern science might have observed, there are sounds that are unmanifest that we cannot experience until we connect our self-discovery process with the self-realization process of the creation and the creator. This is a pursuit beyond just quantified science and hence that is the subject of musical science, which continues in the realm of 'experience' beyond simple quantification.

According to the Indian musical science Anahat Naad = Supreme Brahman himself and that subject is beyond the scope of modern science. Present Author is sure that with further advances in Science, we will in future continue our scientific curiosity to measure the Anahat Naad. Nevertheless, here Science will never achieve full measurement because Anahat Naad is the Naad of Creation.

To fully understand and 'perhaps' experience Anahat Naad one must connect with the Creator himself and that pursuit is where Music excels in its approach. Music spans the disciplines of Science, Religion, Philosophy, and Art and provides us a holistic approach to understanding this Anahat Naad from multiple dimensions.

The following detailed sections exploring the Physiological and Psychosomatic roots of our music from ancient Sangeet Ratnakar treatise are of tremendous value in understanding the genesis of our music. This discussion is a much-needed continuation of the observations on neurology and motor skills covered earlier in the author's 'Science of Rhythm'. The simple act of singing or playing a Rhythmic Instrument or a Flute is not so "SIMPLE" when we understand the many layers of spiritual, scientific, and physical actions happening simultaneously.

Physiological Dimensions

We covered the Metaphysical aspects of the sources and processes of Naad creation in the previous section, now we are diving deeper into the physically manifest form of Naad and its interaction with our human form. As such we now analyze the Physiological evolutionary aspects of the *human body which essentially becomes the first order 'instrument' of experiencing and producing Naad in the material world.*

Order of our Physiological evolutionary analysis is as follows;

1. Human birth (Eternal nature of Soul and Perishable Gross Body)

2. Six substances (Bhava) of human physical form

3. Five Great Elements (Panch Mahabhoota) of Human body

4. Threefold types of Human body constitution

5. Naad related Human organs

(vii) भौतिकदेहभेदाः

स्वेदोद्भेदजराव्वण्डहेतुभेदाच्चतुर्विधम् ।
देहं यूकाऽऽदिनः स्वेदादुद्भेदात्तु लताऽऽदिनः ॥१६॥
जरायोर्मानुषादीनामण्डात्तु बिहगादिनः ।

(viii) मानुषदेहाभिधाने हेतुः

तत्र नादोपयोगित्वान्मानुषं देहमुच्यते ॥१७॥

The physically manifested bodies are of four types based on the nature of its cause of birth. They are those living beings born through the means of Sweat, Sprout, Egg, and Womb. Of these, only the last physically manifest form that is born out of Womb (i.e. Human body) is of interest in our book. Human body and its physical form relevant to our Naad analysis of Science of Melody are to be understood. (16-17)

Lord Brahma is also called by name 'Viraj' in Vedas meaning simply the one who is Luminous. In the first instance of this creation, Brahman itself creates the five

(ix) जीवस्य गर्भाशयेऽवतरणम्

क्षेत्रज्ञः स्थित आकाश आकाशाद्वायुमागतः ।
वायोर्धूमं ततश्चाभ्रमभ्रान्मेघे स्वतिष्ठते ॥१८॥
आहृत्या ऽऽप्यायितो प्रस्तरसो ग्रीष्मे च भानुभिः ।
भानुर्मेघे घनरसं निधत्ते तं वलाहकः ॥१९॥
यदा वर्षति वर्षेण सह जीवस्तदा भुवः ।
वनस्पत्योषधीर्जाताः संक्रामत्यविलक्षितः ॥२०॥
ताभ्यो ऽन्नं जातमन्नं तत्पुरुषैः शुक्लतां गतम् ।
शुद्धान्तबार्याः योषाया निषिक्तं स्मरमन्दिरे ॥२१॥
सहार्तवेन शुद्धं वेद्गर्भाशयगतं भवेत् ।
जीवकर्मप्रेरितं तद गर्भमारभते तदा ॥२२॥

This Human form of ours which is intricately connected to the production and enjoyment of Melody and Naad is first conceived when the 'Jiva' soul descends into the womb of a mother. Here we describe the entire process of the soul's natural descent into a mother's womb.

The Supreme Brahman is the intelligence in all the living organisms and as such when a 'Jiva' descends into the womb, it is this intelligence, which is descending into its field of consciousness. We must remember that the soul 'Jiva' is eternal and non-perishable through association because it IS the part 'Ansha' of the Supreme Intelligence itself. However, the physically manifest form that will evolve through the womb is perishable.

What is the systematic process of descent? We recall that when the physically manifest natural world is created the five great elements are all transformed in the order as follows; from ether to air, from air to fire (as associated with smoke below), from fire to water (as associated with clouds) and finally from water to earth (through vegetation below). This same sequence repeats in the process of descent of the soul into the womb as illustrated herein.

Inside ether (Akasha) resides the all-permeating self-intelligent Supreme who then descends into the air. From the air, it takes on the form of smoke through the heat of light energy from the sun, from the smoke it initiates cloud formation and transforms into clouds. The sun surrounds itself by vapors of steam drawn to it by the heat of its rays. It thus infuses and confines the water of the oceans in those clouds. When the clouds release the waters through the rains, the self-conscious intelligence of Supreme descends with the showers and transfers itself imperceptibly to the earthly growth of vegetation. This then becomes food which when eaten by humans transfers from being subtle to gross in its form as part of the reproductive systems of humans. Soul thus when arrived in the womb, is self-intelligent and self-guided. Through the accumulated residual actions 'Karma' of past lives of the particular individual 'Jiva' soul then accordingly starts the process of physiological manifestation and development into the womb. (18-22)

Thus descends the soul into a womb. We must note that the soul is the subtle body. Soul 'Jiva' has no form, but the steps of transformation that it must go through as described above are essential for it to create its identity by obtaining forms and characteristics of the physical vehicles of its descent into the womb. This entire process is the evolution of manifest reality from its subtle form to its gross form. (Note that sound has two forms audible gross form and subtle which is non-audible. These will be discussed later as Aahat Naad and Anahat Naad respectively)

Physiological Genesis

Stages relevant Music and Naad

Let us now summarize the key stages in the physiological evolution of human form in the womb from embryo to the physically born body and understand which stage are most relevant from a musical Naad perspective. Here we must note that the soul is composed of five great elements in its subtlest of forms and hence the soul has special instincts and abilities to interact with Naad which is in its subtle form (Anahat Naad). However, these abilities are latent and never manifested in general. In the process of human birth, in the few months in the initial phase when the soul descends into the womb, the embryo has all self-intelligence retained still to connect with the subtle Naad.

Then as the Gross from of physical organs develops the subtle intelligence recedes and the Gross intelligence of the physiological body develops and then the Naad relevance shifts from subtle to Gross (Aahat Naad).

i.   First Month – (Subtle Naad Relevance) Self-intelligent and self-conscious soul 'Jiva' which is part of the Supreme consciousness having descended into the womb, transforms itself into a gelatinous substance.

ii.   Second Month – (Subtle Naad Relevance) the embryo gelatinous substance hardens to determine shape relevant to the sex of the human form to be born.

iii.   Third Month – (Subtle Naad Relevance) Five organic seedling forms of head, feet, and hands take shape. All physical growth is imperceptibly slow, uniform and initiated in the third month itself (except for beard, teeth and few other elements that grow post birth)

iv.   Fourth month (Gross Naad relevant) – Organs are distinctly formed and so are the "Bhava" or the latent emotional intelligence of the 'Jiva'. *This marks the first stage in the developmental sequence of the human form where it is musical 'Audible' Naad and it's emotional 'Bhavatmak' tendencies start to develop.*

a. The other thing that happens in the fourth month is that the heart of the embryo is finally taking concrete form and as such is a product of its mother's heart. It starts to 'feel' shared tendencies, feelings, and passions as the mother. Essentially things that the mother does now affect the evolving fetus and the child developing from it. So the likes and dislikes of the mother must be carefully tended to henceforth. The fourth month activates the conscious craving for enjoyment in the embryo shared with the mother.

b. If a mother makes a conscientious effort to listen to soothing and peaceful or devotional music at this stage, similar feelings are produced in the child's musical senses.

c. If starting from the fourth month the pregnant mother wishes to visit temples, the child is bound to have devotional leanings, if the mother wants to visit royal surroundings, the child would lean to have a propensity for wealth attraction, etc. Thus the desires of mothers' heart directly affect the evolution of the 'Jiva's physical form and its leanings in future post-birth.

v. Fifth month (Gross Naad Relevant) – Full consciousness awakens in the fifth month. All matters that relate to emotional and spiritual development indicated in the fourth moth develop fully now. *This Naad mental awareness continues to develop in all subsequent months. Mental awareness is activated.*

vi.   Sixth Month (Gross Naad Relevant) – Bones, Muscles, and Hair and nails develop. *The special Aura 'Ojas' of intelligence and vital force unique to that child also evolves now. Intelligence is activated.*

vii.   Seventh Month (Gross Naad Relevant) – Organs completely develop now. The embryo is self-absorbed now in confined space with fully developed ear cavities that are covered by its hands recalls the past lives and its accumulated actions and torments and contemplates being free from the confinement of the womb. *The embryo can hear actively now outside audible sounds which it could only feel through emotional and other means earlier.*

viii.   The eighth month – (Gross Naad Relevant) Skin and memory are fully developed now. The Vital force of life 'Ojas' is still not stable, it fluctuates from mother to child and if the vital force link is broken, the childbirth is affected negatively. Ojas regulates the flow of vital life force through the heart and onwards to other organs. *Memory is activated.*

ix.   Ninth Month – (Gross Naad Relevant) Here, the embryo is fully developed, the vital force is stabilized, and Baby is ready to be born either in the natural course now or in the next 10th or 11th month in rare cases.

Six Substances ('Shad Bhava') Embryo.

Let us also quickly understand the Six Bhava of the various parts and elements that support the embryo from genesis until birth. We discussed the subtle and gross form of intelligence as the embryo is evolving and these correspond to the Naad and its recognition by the soul. The second factor that also affects the Naad relevant development in the yet to be born child is the composition of the six substances of the matter that embryo is formed of.

i.   Matraja – derived from mother (These are the material substances and emotional qualities as derived from mother.

ii. Pitraja – derived from father

iii. Rasaja – derived from "Rasa" or serum in physiological context (This substance is more related to the Gross audible form of Naad)

iv. Atmaja – derived from the transmigratory soul (This substance is more in tune with the Subtle form of Naad)

v.   Sattva – derived from and pertaining to the pure sense of mind

vi. Satmyaja – acquired by adaptation and habit (genetic hereditary physiological adaptation). The embryo inherits these qualities from the ancestors and lineage. This is Naad relevant in that if the ancestors had acquired musical sense and honed the skills of melody and rhythm, through subtle memory and intelligence from this substance, this knowledge can pass unto the new embryo.

The entire composition of the embryo and its evolutionary physiological form is derived from these six substances as inherited by the fetus of that 'Jiva' soul. This summarizes the complete physiological process of human body development from genesis to Birth.

There are many stages where the future music and Naad relations of this 'Jiva' are impacted while it is evolving into a full physical form. These six substances (Shad Bhava) are the extra layer of all accumulated intelligence and musical memories of past lives. The Subtle body of which the soul is constituted is eternal and it has a subtle memory of its past that it carries forward with the gross physiological developments taking form above. It is therefore wisely acknowledged by the Indian musical sages that the musical Naad 'yatra' Journey is not confined to one life but spans across many lives and caries forward Soul's continuing and evolving efforts positively into the future lives.

Five Great *Elements (Panch Mahabhuta)*

After covering the soul and its birth after taking the form of a physical body we also discussed six types of substances (Shad Bhava) of emotional and subtle past life intelligence that are passed on to the new body. Now let us study the five great elements of nature that we have discussed earlier and understand how they manifest in the human form.

We recall that everything in nature is unmanifest initially in the Supreme Brahman. When the world manifests, the five material elements are the manifestations of Brahman and the permutations and combinations of these five cause the entire universe to evolve. The human body is but one small piece of this creation. As such, the human body is also composed of specific contents of the 'Panch Mahabhuta' and acquires their qualities described later.

The "macrocosm" consists of six elements the great five material elements plus the eternal Consciousness of the Brahman. This sixth eternal consciousness is the Spiritual element that we know as 'Spirit'. These same six compose the "microcosm" of our body, which in Vedas is also addressed as the 'Purusha'. This Purusha is said to consist of the earth as its image, water forms the liquids, fire constitutes the heat, air acts as vital breath, the expanse of hollowness is the Ether and the sixth element of Spirit is the Brahman itself as the inner soul 'Jiva'.

These five Great elements are directly corresponding and related to our five sense objects. Sense organs are related to their corresponding objects of perception. These objects of perception are nothing but the five elements in their various combinations. Thus even though related we speak in common terms either of the sense objects or the materials. E.g., the object of Sound is known to be a sense object i.e. it defines a quality or object of one of our senses (in this case hearing). And this sense object of sound in our body is indicative of the great element of Ether 'Akasha'. Senses and the Sense Objects are related through their respective 'Tanmatra'. The table below illustrates the complete summary of the Panch Mahabhuta and their corresponding functions in our Body. Of note is that Prana is essential for music and Naad creation and so is Ether.

Sr.	Great Element	Sanskrit Names	Qualities Sense Objects	Sub class	Notes
1	Ether	Akasha	Sound		
			Faculty of Hearing		
			Porosity		
			Differentiation		
			Subtle Intelligence		
			Hollowness from space		
2	Air	Vayu	Touch		
			Sense Organs for Touch		
			Five types of motions	Upwards	
				Downwards	
				Contraction	
				Linear	
				Expansion	
			Ten Modifications of Air in body.	Prana	Stationed below root of navel and operates through mouth, nostrils, navel and heart causes verbalization of speech, inhaling and exhaling, as well as sneezing and coughing.
				Apana	Resides in the anal region, Waist, legs, root of navel, thighs and knees and is responsible for impurity discharge
				Vyana	Dwells in eyes, ears ankles, waist and nose. Draws in breath and holds it in and discharges breath.
				Samana	Runs through whole body and its 72,000 nerve channels (nadis). This air takes the digestive (fire) and caries the nutrition from lymph-chyle (rasa) of food and drink and distributes it to tissues and muscles proportionately.
				Udana	abides in hands, feet and joints of limbs. It lifts body up and also is responsible for last breath of life
				Naga	resides in skin
				Kurma	resides in blood
				Kykara	resides in flesh
				Devdatta	resides in fat
				Dhananjay	resides in bones
			Roughness		
			Lightness		

Sr.	Great Element	Sanskrit Names	Qualities Sense Objects	Sub class	Notes
3	Fire	Agni	Sight		
			Form		
			Bile		
			Digestion		
			Luster		
			Wrath		
			Sharpness		
			Heat		
			Vigor		
			Splendor		
			Valor		
			Intellectual Essence		
4	Water	Jala	Sense of Taste		
			Relish		
			Coolness		
			Viscidity		Sweetness of speech
			Fluidity		
			Perspiration		
			Urine		
			Softness		
5	Earth	Prithvi	Sense of Smell		
			Odor		
			Stability		
			Fortitude		
			Heaviness		
			Beard		
			Hair		
			Nails		
			Bone		
			Hard materials		

Seven Supportive Tissues and Receptacles

We now cover in summary a few topics with the goal of providing a complete understanding of our Human Body, which is the instrument of Naad production first amongst all other instruments. However, these subsequent topics would deviate us too much from our subject which is music and hence we keep the introduction brief.

The human form has seven skin layers and seven membranes of which the exterior of the human form is composed of according to the Indian biological treatise from ancient times. In between these skin and membrane, layers in various folds lie the seven key supportive tissues (Dhatus) of the human form listed as;

I.	Serum
II.	Blood
III.	Flesh
IV.	Fat
V.	Bone
VI.	Marrow
VII.	Semen

Matched with these seven tissues are seven receptacles that act as containers (Ashayas) for things such as blood, bile, undigested food, wind, urine, etc.

Threefold types of Bodies

135

There are three main types of bodies and this threefold classification derives from the permutation and combination of the differing degrees of the five great elements (Panch Mahabhut) in the body along with the seven supportive tissues (Dhatus). Based on this the types of human body analysis can be grouped into a threefold group.

a) The medical constitution of the body consists of wind, bile, and phlegm (vata, pita and kafa).

b) The physical constitution consists of five great elements.

c) Thirdly, metaphysical constitutions of the body consist of the three subtle qualities of the soul that come from the embryo's inception... i.e. Rajas, Tamas, and Satvik.

Each of these bodies is formed of six types of organs and the sub-organs that comprise of the Dhatus described above. There are precise numerical and qualitative identifications of each of the other aspects of the human body such as types of joints, bones, canals, muscles, etc. A discussion of these elements is avoided here to deviate from our main topic of Science of Melody. However, for the reader interested in learning more about the depths of Human biological science and its Indian perspectives volumes from Charaka and Shushruta two great Ayurveda experts from ancient times are suggested.

Psychosomatic Dimensions

We have so far covered the entire process of the physiological evolution of our human form from the inception of the body to its relevant specifics of the constitution. We now explore the equally interesting and often unexplored dimensions of psychosomatic analysis of our Human body and its relevance in music production.

Since music universally is understood to have a spiritual dimension, we must explore these aspects of Indian musical science in further detail. The analysis must be conducted by a genuinely curious practitioner of Science of Melody and Rhythm as to what is the context of the yogic knowledge of topics such as human intelligence and consciousness and their relation to Naad. What are the physical and inner centers of spiritual powers in our body that directly influence our music production and enjoyment?

Intelligence and Consciousness

The human heart is the starting point of our psychosomatic understanding of our body and its relation to Naad. The human heart is hollow and in the shape of an inverted lotus between the liver and spleen. Most importantly, from our point of view, the heart controls the brain and per the Indian ancient sages, the heart is the epicenter of our human consciousness. Within the heart resides the human consciousness that is also connected in the subtle form to the universal consciousness of the Brahman in a latent form.

The human form is comprised of spirit-soul and body. Thus, we must understand their interplay and relationship further. The Spirit 'Atma' is the part of the Supreme Brahman as we discussed earlier and is the 'Divine' element of the human form. The soul is the 'Jivatma' the embodied self that is eternal and trans-migrates between bodies. The body has the human heart described above approximately in inverted lotus format and is the seat of the spirit-soul consciousness.

This consciousness has three states; the waking stage (visva), the dreamful rest stage (taijasa) and the dreamless sleep state (prajna). In modern psychology, these three states closely correspond to conscious, subconscious and unconscious. The third term unconscious is not fully applicable to our third stage as we describe below.

I.   'Visva' Conscious (Waking) stage – This literally is the stage when we are awake and interacting with the world around us, hence the name 'visva' meaning world. The key aspect to understand here is that the spirit 'Atma' itself sleeps when it is covered by ignorance and darkness and the heart's lotus petals close and enclose the spirit. When the heart opens up with intelligence and light, the spirit awakens and the self-conscious soul 'Jiva' enters the waking state.

II.   'Taijasa' Dreamful Sleep Subconscious stage – When external senses are withdrawn into the heart for rest but the mind remains awake, this state is activated. Part of the brain and the thought process is actively engaged and this results in visions and dreams while sleeping. When sense perception is absent as the senses are withdrawn inside the heart, the thoughts in our mind remain as the only objects of consciousness. Hence, in this state, the distinction between the ideal and actual form of the objects is lost. Thus, dreams are as real in the dreamful state as are the physical objects in the waking state.

III.   'Prajna' Dreamless deep sleep stage – This stage is caused by the mind being withdrawn into the vital force 'Prana'. The 'Prana' is the only consciousness active in this stage and is the only thing that ensures the 'alive' nature of human form while in this stage. This is also the stage when the human soul is closest to its subtle consciousness with the Supreme Brahman. In the other two stages, the 'Self-consciousness' of the individual is still active and acts as a veil of ignorance hiding the soul's true relation to the Supreme and hence in this 'Prajna' state there many spiritual possibilities of divine enlightenment. Thus, we can see that it would be inaccurate to call this stage unconscious corresponding to modern psychology.

This Psychosomatic dimension is further explored in depth in the next topic that due to its importance necessitates a separate section for itself. We now initiate the analysis of the yogic science of inner chakras in our human form. The consciousness and intelligence that is activated by practicing and understanding these concepts are essential in producing Melody and music of the highest spiritual order.

Yoga Chakras and Melody

Major Chakras and Nadis
for Kundalini Risings

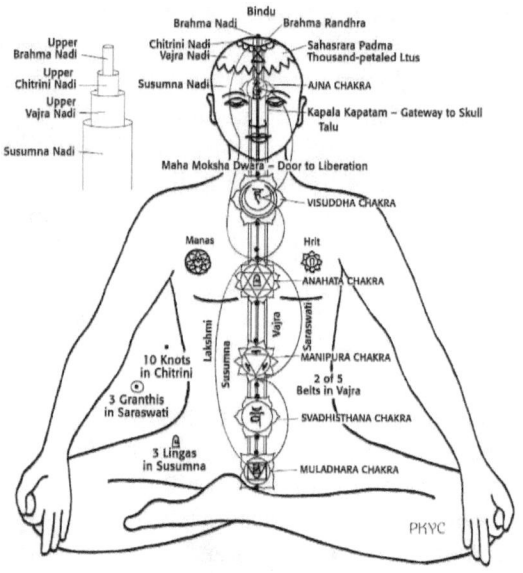

Six main psychosomatic centers exist inside our body as identified by ancient Indian yogis and sages of yore. With their deep knowledge of Vedas and practical application in the human physical form, the ancients knew of techniques to understand our relation to the whole Brahman in a complete manner. Knowledge of these helps the spiritual musical soul in producing the complete Naad.

There are many branches of Yoga in the Indian Vedic system for embarking on a journey to understand and unite with the Supreme Creator. All these systems vary in their approaches and practice in some form or another. However, the common denominator across all the Yoga systems is human physiology and their identification of its inner power centers depicted above which we will explore in detail in this section. These individual power centers situated physiologically in alignment with the human spinal column act as special points at various levels of human consciousness. The knowledge and control over these special points bestow upon the person several psychosomatic skills that otherwise might remain untapped.

1 Muladhara Kundalini (Root)

घ. हठयोगानुसारि निरूपणम्

(i) दशचक्राणि

1. आधार-चक्रं कुण्डलिनी च

गुदलिङ्गान्तरे   चक्रमाधाराख्यं   चतुर्दलम् ।
परमः   सहजस्तद्वदानन्दो   वीरपूर्वकः ॥१२०॥
योगानन्दश्च   तत्र   स्यादीशानादिदले   फलम् ।
अस्ति   कुण्डलिनी   ब्रह्मशक्तिराधारपङ्कुजे ॥१२१॥
आब्रह्मरन्ध्रमृजुतां   नीतेयममृतप्रदा ।

D. Genesis : The psychophysical viewpoint.

(i) Ten *cakra-s* (psychophysical centres): 120-145b

The fundamental human power and psyche center is located in between the anus and the genital area and its name is 'Muladhara Chakra' (Fundamental Foundational Cycle). This is in the shape of a four-petalled lotus.

The four petals represent the four cardinal directions and their intermediate points, 'Ishan' petal (Northeast) corresponding to Supreme Bliss, 'Agneya' petal (Southeast) corresponding to spontaneous happiness, 'Nairutya' (Southwest) corresponding to Heroic Joy and 'Vayavya' (Northwest) corresponding to Divine Unity.

In its center lies the creative power of the universe the creative spirit of the Supreme Brahman. This is the 'Kundalini', which remains coiled in its natural state and when unfolded inside the Muladhara chakra bestows upon the 'Jiva' the Supreme Liberation 'Moksha', and Enlightenment. (120-121)

In order to understand the complete spiritual and psychosomatic sensibilities of our music, we must go to the root of the source of all life and all music **'inside'** our body. This is the Brahman as we have already discussed earlier. The 'Jiva' or the embodied soul accepts a physical form during the process of birth as we have seen. During the process of birth, this 'transmigratory embodied consciousness of the soul' establishes its seat in Kundalini which is the 3.5 turn coiled 'Potential Energy' of the universe with its corresponding supreme consciousness. It can also be said to be the 'residual' power center whereby the transmigratory soul, once having decided to take on a specific physical form undergoes the process of birth and stores the residual supreme consciousness which is the Brahman itself hidden in Kundalini away from the material world and material plane.

This energy resides here in its 'Potential' or steady form. The Kundalini is enclosed and protected in the Muladhara Chakra and it is through the Meditative practice of Pranayama and Yoga that one can activate this supreme energy and convert it into its true 'Kinetic' creative nature. When the Kundalini is 'awakened' it transforms from being static in its potential form to active in its kinetic form. Through spiritual meditation, this kinetic energy moves upstream with tremendous force through the spinal column, specifically the 'Sushumna' central energy channel of life force in the spine.

Once activated one can enable its journey upstream towards our heart and brainpower centers ultimately reaching the upper cerebrum and the spiritual center of the cerebrum, the 'Brahmarandhra'. Finally when energy that was 'awakened' in Kundalini reaches the spiritual center of consciousness in the brain the Yogi attains 'enlightenment' and experience equivalent to timelessness and freedom from birth and death cycles 'Moksha'.

The energy reaches on activation the next Chakra upstream and the Yogi finds him/herself at the next plane of consciousness and spiritually enlightened value system. The subject of Kundalini Yoga specifically describes techniques of meditating on each of the petals of Chakras being discussed here. Each lotus petal has a specific seed sound 'mantra' attached to it which when leveraged with meditative concentration enables one to activate spiritual energy that remained untapped till that point. Further Kundalini Yoga details would deviate us beyond our scope of this book hence we present this summary discussion for our complete understanding of the Chakras and their intimate relation to our spiritual and physical body systems. For the curious seeker, the 'sounds' of the letters corresponding to each of the petals are documented herein.

Six primary Chakra petals corresponding Sanskrit letter sounds						
Lotus Petals	Muladhara 4 Petals	Swadhishtan 6 Petals	Manipura 10 Petals	Anahat 12 Petals	Vishuddhi 16 Petals	Aagya 2 Petals
1	va	ba	da	ka	sixteen	ha
2	sa	bha	dha	kha	sanskrit	ksha
3	sha	ma	na	ga	vowels	
4	sa	ya	ta	gha	from	
5		ra	tha	na	a	
6		la	da	ca	to	
7			dha	cha	ah	
8			na	ja		
9			pa	jha		
10			pha	jna		
11				ta		
12				tha		

2 Swadhishtan Chakra (Self-Support)

2. स्वाधिष्ठान-चक्रम्

स्वाधिष्ठानं लिङ्गमूले षट्पत्रं चक्रमस्य च ॥१२२॥
पूर्वादिषु दलेष्वाहु: फलान्येतान्यनुक्रमात् ।
प्रश्रय: क्रूरता गर्वनाशो मूर्च्छा तत: परम् ॥१२३॥
अवज्ञा स्यादविश्वास: कामशक्तेरिदं गृहम् ।

The second Chakra that is residing upstream from the previous Chakra is the 'Swadhishtan' Chakra, literally which means, "self-abiding or self-supporting cycle". This Chakra resides at the root of the human genital area inside the spinal energy column. This is a six-petal lotus form chakra and meditating upon it develops certain specific psychological qualities through the power of concentration. 'Swadhishtan' is also the location of the seat of our passion and desires known as 'Kama' (122-123)

We saw that the creative power of the Supreme Brahman resides at the very root of our Spinal column inside the main central life force energy channel called 'Sushumna'. This very bottom root is the 'Muladhara' Chakra. Enclosed inside its four petals resides the 3.5 coiled serpent format of Supreme Kinetic Energy of the Creator in the form of 'Kundalini'. Its natural state is to remain coiled and static just like a baby lying in the womb active consciously but static physically coiled up.

When through meditative concentration this 'Kundalini' is 'awakened' it uncoils and releases the energy that remained static so far with tremendous force pushing it into this 'Swadhisthan' chakra. The word Swadhishtan etymologically is broken down as Swa + Adhi + Sthan per Sanskrit root meanings and literally translates appropriately as "One's Own Seat". Meaning this is the self-supporting seat of one's own energy as it arises from Muladhara (Kundalini) and resides here first.

This spiritual center of 'Swadhishtan' controls all the main self-centered pursuits of our being. These essentially are the five main roots of all selfish, material and sinful activities that plague our existence. These are;

1. Kama (Desire)
2. Krodha (Anger)
3. Lobha (Greed)
4. Moha (Delusion)
5. Ahankara (Ego)

Each of these are the 'enemies' of our spiritual enlightenment and true selfless progress to connect with the Universal Naad Brahman. These each correspond to the five of the six petals of the 'Swadhishtan' chakra. The sixth easternmost petal is that of "courtesy" or 'Vivek' in Sanskrit which enables one to understand the evil nature of the five enemies and act to control them. Meditation on this self-supporting chakra of 'Swadhishtan' thus enables one to activate its energies and conquer the five enemies of our selfish material existence with the right 'courtesy' and migrate our consciousness upstream to next plane.

When Kundalini is released from this second Chakra and through deliberate and concerted meditative efforts is pushed up to the third chakra, the Yogi then attains complete self-control over the mental states arising out of the influence of the five evils mentioned above. The

Yogi then is said to have conquered their Kama, Krodha, Lobha, Moha, and Ahankar.

3 Manipur Chakra (Navel)

3. मणिपूर-चक्रम्

नाभौ   दशवलं   चक्रं   मणिपूरकसंज्ञितम् ॥१२४॥
सुषुप्तिरत्र   तृष्णा   स्याबीर्ष्या पिशुनता तथा ।
लज्जा भयं घृणा मोह: कषायो ऽथ विषादिता ॥१२५॥
क्रमात्पूर्वादिपत्रे   तु   स्याङ्ङानुभवनं   च   तत् ।

This is the third Chakra resides upstream from previous and is called the 'Manipur' chakra. This is the ten petalled lotus format psychosomatic power center. This chakra resides at the root of the navel in our solar plexus. It is the location of the special type of 'Prana' (Lifeforce) called 'Bhanu' literally meaning Sun. Meditating upon this center allows one to control their mental states related to; dreamless deep sleep, cravings, jealousy, looking at others' faults, bashfulness, fear, hatred, ignorance, impropriety, and dejection. (124-126)

Just as the sun is illuminated with its brilliance 'Tejas' so is this chakra illuminated with powerful brilliance due to this prana of 'Bhanu' residing there. The brilliance is symbolically akin to a jewel 'Mani' and hence the name of the chakra is the one where the jewel resides. 'Manipur'. This is the chakra also called as 'Nabhi Padma' as it is the lotus petalled flower-shaped residing at the root of our navel 'Nabhi'. Solar plexus is the center of our body and of great prominence physiologically. This is the location of this chakra and the two main sympathetic chains of physiological flows are centered here as well. These are the right sympathetic chain (Pingala Nadi) and the left sympathetic chain (Ida Nadi). The 'Ida' which is to the left of the central life force channel 'Sushumna' controls the feminine aspect of human nature and the creative right side of our cerebrum. 'Pingala' which is to the right of the central channel controls masculine aspects of our nature and the left analytical side of our cerebrum. 'Manipur' chakra is where these Ida and Pingala intersect and also form the center of their move upwards ultimately joining in our cerebrum at the 'Agya Chakra' which we will cover later.

The material physical 'gross' existence of our body is directly related to the powers concentrated in the three chakras discussed so far starting from the lowest at the root of spinal column 'Muladhara' followed by 'Swadhishtana' and then this 'Manipur' chakra. The higher chakras above Manipur are more concerned with psychological development and have more direct prominence in connecting us with the spiritual aspects of Naad. The physical aspects of our body and its 'clean' functioning are better controlled by these three lower chakras mentioned so far. In other words, first after cleaning the physical 'gross' aspects of our existence, we then can move up to connecting with 'spiritual' and 'subtle' aspects of our soul and the Naad Brahman.

4 Anahat Chakra (Heart)

4. अनाहत-चक्रम्

हृदये ज्नाहतं चक्रं शिवस्य प्रणवाकृते: ॥१२६॥
पूजास्थानं तदिच्छन्ति दलैर्द्वादशभिर्युतम् ।
लौल्यप्रणाश: प्रकटो वितर्को ज्प्यनुतापिता ॥१२७॥
आशा प्रकाशश्चिन्ता च समीहा समता तत: ।
क्रमेण दम्भो वंकल्यं विवेको ज्हंकृतिस्तथा ॥१२८॥
फलान्येतानि पूर्वादिदलस्थस्यात्मनो जगु: ।

In the heart is situated the fourth chakra whose name is 'Anahat' chakra. It represents the 'unmanifested' power of the Supreme Brahma and is in the form of a 12 petalled lotus. This chakra lotus is the place of worship of 'Pranav', which is also knowns as 'Om' and is another name for Supreme Brahman in the form of Lord Shiva.

The fruits of meditating upon this chakra starting on each petal from its easternmost petal are positive aspects achieved through the conquering of fickleness, clear thoughts, repentance, hope, and light of knowledge, freedom from worry, freedom from evil, equanimity, vanity, mental instability, discernment and will power. (127-128)

This is the chakra which has direct relevance now to our interest in understanding and mastering the Naad yoga through the practice of music. This is the seat of the "Paramatma". Hence understanding this chakra is of paramount importance for the practitioner of Rhythm and Melody. The word 'Anahat' literally means 'inaudible and unmanifested'. We have covered the Naad of two types and their processes of evolution in other relevant chapters and it would be ideal to refresh here that the 'Anahat' Naad means the sound that is essential for sustainment of the creation and our universe and is an unmanifested form of Supreme Brahman itself. This same 'Anahat' Naad initiates its journey of manifestation in our physical realm starting from this chakra and hence the name 'Anahat' chakra.

Our heart is the seat of our physiological control and operating unit and within the heart the 'Anahat' chakra is the seat of embodied consciousness that we also know as 'Jivatma'. Anahat Naad is the unmanifested sound of the Brahman itself and is the Supreme vibrating pulse of all creation and its cause. This Anahat Naad is not heard in our physical realm, however, in the chakra corresponding to its name in the heart, the Supreme Brahman manifests itself as 'Shabda Brahman' which is essentially the first step towards the living creation experiencing the physical vibrations and manifestations of the sound through the concept of Shad Brahman. The Shabda Brahman is nothing but the Supreme Brahman manifested in the form of an inarticulate word with audible sound. All creation of the universe starts from this first audible sound and that is known by its other names as 'Pranav' or 'Om'.

'Om' is also called 'Omkar' and in English, it is spelled as Aum with three letters and two vowels. 'a', 'u' and consonant 'm'. There is also a dot written over the Sanskrit alphabet of 'Om' and this is known as the 'Chandarbindu'. The divine spiritual meaning of these three letters broken down into the sound of Aum with the dot represents the four states respectively;

1. 'A' - Waking state

2. 'U' – Dream state

3. 'M' – Dreamless sleep state &

4. '~' – 'Turiya' state of Total Complete oneness with Creator represented by the dot above the words.

It is this 'Om' which is the manifestation of Supreme Brahman in the form of 'Shabda Brahman' and its seat is in the Anahat chakra. Every sound of speech, language and divine creation starts from this manifestation in Anahat chakra. There would be no melody and no music were it not for 'Omkar' hence meditating and knowing it through the exploration of this chakra is essential for the musician who wishes to become one with the 'spirit' of music. He who understands this chakra and its functioning attains full mastery over all speech and through this can create and destroy.

5 Vishuddhi Chakra (Throat)

5. विशुद्धि-चक्रम्

कण्ठे ऽस्ति भारतोस्थानं विशुद्धिः षोडशच्छदम् ॥१२९॥

तत्र प्रणव उद्गीथो हुंफड् वषडथ स्वधा ।

स्वाहा नमो ऽमृतं सप्त स्वराः षड्जादयो विषम् ॥१३०॥

इति पूर्वादिपत्रस्थे फलान्यात्मनि षोडश ।

'Vishuddhi Chakra' is the fifth in the series of chakras moving upwards from the bottom one (Muladhara). This chakra resides in the throat and larynx and is the psychosomatic center of the 'purity cycle'. This chakra is the abode of goddess 'Bharati' who is the presiding deity over all knowledge and this chakra consists of 16 petalled lotus.

Contemplating upon these sixteen petals in sequence from easternmost provides the yogi with control over sixteen skills 'siddhis' which are listed in sequential order: Pranav (omkar), Udgith, humphat, vasat, svadha, svaha, namah, nectar (amrit) seven swara of music (seven tones) and vish (poison). (129-130)

This the second chakra that directly connects our musical linkages from the physical world to the spiritual and psychosomatic world. This chakra and its exploration provide mastery over the seven Swara (tones) of Melody in its universal format. The 'Shabda

Brahman' manifests itself inside the Anahat chakra and then moves upwards into this Vishuddhi (purity) chakra.

Once the embodied consciousness (our ego-soul) known also as 'Jivatma' finds and identifies its eternal connection with its Supreme Brahman consciousness (boundless eternal subtle soul) knows also as 'ParamAtman' inside the Anahat Chakra, it essentially gets 'Purified' and hence enters next chakra which is called the 'Vishuddhi' (Pure) chakra. Here in this Purity chakra, the 'Jivatma' having understood its root relationship to the eternal soul, attains the power to perceive all three stages of time; viz. past, present, and future. All barriers of ignorance are eradicated upon mastering this chakra and doors of boundless consciousness are opened.

Attaining full knowledge of the 'ParamAtman' i.e. Supreme Brahman through mastery over previous Anahata chakra and then by focusing the mind upon this chakra of 'Vishuddhi' (purity), one becomes a wise sage and enjoys undisturbed peace. Able to see all three stages of time the yogi then lives long and acts to benefit all mankind.

The sounds of the sixteen petals of this chakra are directly related to the Vedic chants. music is derived according to the Vedic traditions from the Sama Veda and its hymns and their chants. The petals of this chakra emphasize this relationship to audible Naad. The previous chakra was related to the unmanifested Naad and hence called 'Anahat' chakra. This chakra here is related to the 'Ahata' Naad (manifested audible sound). Concentrating upon its first petal one gains intimacy with the 'Omkar' itself, the second petal gives one understanding of the 'Udgith'. This Udgith is essentially the second part of the Sama Veda chants. (There are five parts of the Sama Veda chants and they are Prastav, Udgith, Pratihara, Updrava, and Nidhana.)

All Vedic music is said to have evolved from Sama Veda hymns and is essentially part of the "Shabda Brahman. The Seven Swara (Shadaj, Rishabh, Gandhar, Madhyam, Pancham, Dhaivat, and Nishad) are all also derived from these and the focus on this chakra enhances one's mastery over these seven notes of our Melodic music. Thus one attains purity of pronunciation and intimacy with all the melodious 'Ahata' Naad through mastery of this chakra.

6 Aagya Chakra (Third Eye)

' 7. आज्ञा-चक्रम्

भ्रूमध्ये त्रिवलं चक्रमाज्ञासंज्ञं फलानि तु ॥१३३॥
आविर्भावाः सत्त्वरजस्तमसां क्रमतो मताः ।

The sixth main Chakra that is residing between our eyebrows is 'Aagya' or 'Ajna' chakra. It means the chakra cycle of Supreme Command from the Supreme Brahman as the guide. This chakra has three petals and those control the three manifested 'Guna' qualities of the Brahman in the living universe. They are Sattva, Rajas, and Tamas respectively. (133)

The word 'Aagya' in Sanskrit literally means 'command'. Here the commander is the Supreme Brahman he is the penultimate Guru (guide) and the mastery over this chakra establishes a permanent connection between the embodies soul 'Jivatma' and its eternal father or Guru, the supreme creator. The position is exactly in the middle of the eyebrows which is also known by some as the 'spiritual or divine eye' It is in this chakra where the 'Antahkaran' is guided by the Supreme Brahman. Antahkaran is the fourfold composition of mind-stuff (consciousness), intellect, ego and will powers.

These are the main SIX phsychosomatic centers of our physical bodies that connect us directly to our spiritual awareness across many planes of conciousnes. There are also more chakra which are well known but not considered amongst the primary six per the Kundalini Hatha Yoga practices. We list them here for completeness and also because Sharangdev illustrates it in his SangeetRatnakar treatise which forms the foundational basis of our analysis here. Present author also feels that these other chakras and their understanding gives a practioner of music complete understanding of Naad and its flow in our pshycosomatic systems.

7 Kaal (Lalna) Chakra

6. ललना-चक्रम्

ललनाऽऽख्यं घण्टिकायां चक्रं द्वादशपत्रकम् ।।१३१।।
मदो मानस्ततः स्नेहः शोकः खेदश्च लुब्धता ।
अरतिः संभ्रमश्चोर्मिः श्रद्धातोषोपरोधिताः ।।१३२।।
फलानि ललनाचक्रे स्युः पूर्वाविदनेष्विति ।

'Lalna' chakra is 12 petalled and situated in the back of our necks. This chakra is also known as 'Kaal' chakra. It provides command over our material emotional states as we live in this physical world. (131-132)

This seventh chakra is called 'Kaal Chakra' or 'Lalna Chakra' and resides in our root of the palette (back of the neck). Thus it is in between the Vishuddhi chakra at the bottom and Aagya chakra above it. The 'Kaal' is the material aspect of time in which we are existing and hence this chakra per its name is connected to the material aspects of our existence such as six basic urges of life; Hunger, Thirst, sorrow, delusion, decay, and death. These six are basic urges that reside in our vital breath, mind and body and are controlled by obtaining mastery over this Kaal Chakra. Other emotional material qualities of life such as love, haughtiness devotion, and satisfaction, etc. are controlled by this chakra. This is a 12 petalled chakra.

8 Manas Chakra (Mind)

8. मनश्चक्रम्

ततो ड्प्यस्ति मनश्चक्र' षड्दलं तत्फलानि तु ॥१३४॥
स्वप्नो रसोपभोगश्च प्राणं रूपोपलम्भनम् ।
स्पर्शनं शब्दबोधश्च पूर्वादिषु बलेष्विति ॥१३५॥

Higher than the 'Aagya' chakra lies the 'six-petalled 'Manas' (mind) chakra. Meditating upon this area empowers the yogi with control over dreams, taste, smell, visual beauty, touch, and sound. (134-135)

Having covered the six primary yoga chakras earlier, we now also cover the important aspects of mind and body cycles next. IN this present 'Manas' (mind) chakra the psychosomatic cycle essentially deals with the conscious and sub-conscious states of our mind. Six petals of this chakra signify the five senses of our perception and the sixth sense of our intellect. The yogi attains command over their six senses and their perceptions and corresponding mental activity. Sound is one of the perception of our sense of hearing and is linked to our ability to produce and enjoy music and as such this chakra provides us that link.

9 Soma Chakra (Moon)

9. सोम-चक्रम्

ततो ऽपि षोडशदलं सोमचक्रमितोरितम् ।
दलेषु षोडशस्वस्य कलाः षोडश संस्थिताः ॥१३६॥

कृपा क्षमाऽऽर्जवं धैर्यं वैराग्यं धृतिसंमदौ ।
हास्यं रोमाञ्चनिचयो ध्यानाश्रु स्थिरता ततः ॥१३७॥

गाम्भीर्यमुद्यमोऽच्छत्वमौदार्यंकाग्रते क्रमात् ।
फलान्युदयन्ति जीवस्य पूर्वादिदलगामिनः ॥१३८॥

Above the 'Mans' resides the 'Soma' (Moon) chakra cycle. This is sixteen petalled and corresponds to the sixteen phases of the moon and its waxing and waning cycles. Vedic science often refers to Sixteen 'Kala's that derive from the moon's sixteen phases. Emphases on the sixteen petals of this chakra empower the yogi with sixteen essential qualities of an enlightened being viz. grace, forgiveness, straightforwardness, detachment, patience, purity of heart, generosity, etc. (137-138)

'Soma' chakra is residing in our middle cerebrum and is located above the 'Manas' chakra. 'Jivatma' embodied soul through its yogic effort rises above the Aagya (conquering the opposite aspects of our manifested world) and Manas chakra (conquering the world of relative perceptions of senses and desires). Thus it realizes the path of true oneness with the Supreme Brahman and reaches the 'Soma' chakra and attains the coolness and peaceful tranquility just as that of the moon. Moon and its rays are cool and soothing to the soul and its sixteen phases reoccur in a cyclical manner are sources of all the true spiritual and divine sentiments of an enlightened soul.

10 Sahastradhara

10. सहस्रपत्र-चक्रम्

चक्रं   सहस्रपत्रं   तु   ब्रह्मरन्ध्रे   सुधाधरम् ।
तत्सुधासारधाराभिरभिवर्धयते   तनुम्   ॥१३९॥

The 'Sahastradhara' chakra is composed of thousand petals and is situated in the top aperture of our cerebrum. The thousand-fold nectar of timeless consciousness resides here at the point of 'Brahmarandhra', top of our brain (cerebral aperture) and nourishes our body with its life force. (139)

The thousand-petalled chakra of 'Sahastradhara' (thousand streams) is also known as 'Sahastrapatra' (thousand petals) and is located in the spiritual void of our cerebral aperture 'Brahmarandhra'. This void is often termed in Vedic terms as 'Param Vyom' (highest ether), 'Bindu' (void), 'Parabindu' (highest void), 'Shunyabindu' (zero) or 'Ishwara' (Manifested God). Here in this highest ether which is also the subtlest of space resides the light which is at one formless and with form. The light which is eternal and yet unmanifested.

When our consciousness reaches to this highest place, it then transcends time and becomes timeless and can attain any form or no form at will. This is the stage in which many practices and paths call by different names as 'moksha', 'immortality', 'liberation', 'heaven', 'Ambrosia', 'nectar', etc.

11 BrahmaGranthi (Root)

(ii) ब्रह्मग्रन्थि :

आधाराद् द्वचङ्गुलादूर्ध्वं मेहनाद् द्वचङ्गुलादधः ॥१४५॥
एकाङ्गुलं देहमध्यं तप्तजाम्बूनदप्रभम् ।
तत्रास्तेऽग्निशिखा तन्वी चक्रात्तस्मान्नवाङ्गुले ॥१४६॥

देहस्य कन्दोऽस्त्युत्सेधायामाभ्यां चतुरङ्गुलः ।
ब्रह्मग्रन्थिरिति प्रोक्तं तस्य नाम पुरातनैः ॥१४७॥
तन्मध्ये नाभिचक्रं तु द्वादशारमवस्थितम् ।
लूतेव तन्तुजालस्था तत्र जीवो भ्रमत्ययम् ॥१४८॥

'Brahmagranthi' is the cycle of umbilicus also known as 'nabhi chakra'. It has 12 spokes and like a spider residing within the center of its own web, the embodied soul 'Jivatma' resides there in the center of the Brahmagranthi.

Its exact co-ordinates are described as follows. Center of our body is situated two fingers length above the 'Muladhara chakra' and two fingers breadth below the genitals. It takes up space of one finger breadth and within that space resides the center of the body shining like smelted gold. This is the spiritual solar plexus.

Nine fingers from this center of the body is situated a flame of fire (life force). The ancients and the Vedas call this place 'Brahmagranthi', which is four fingers wide and four fingers in length and within it resides the 'soul'. (145-148)

This is the very 'Root' of our physical existence. The very place from where we came from and residence of our embodied soul. Meditating upon this and understanding this location is useful for those who care to explore 'Jivatma's eternal past, present, and future.

12 Sushumna Ida Pingala (Nadis)

(iii) सुषुम्णा, नाड्यन्तराणि च

सुषुम्णया   ब्रह्मरन्ध्रमारोहत्यवरोहति ।
जीव: प्राणसमारूढो रज्ज्वां कोल्लाटिको यथा ॥१४९॥

सुषुम्णां परितो   नाड्यः: कन्दादाब्रह्मरन्ध्रतः ।
कन्दीकृत्य स्थिता: कन्दं शाखाभिस्तन्वते तनुम् ॥१५०॥

ताभ्र   भूरितरास्तासु मुख्या: प्रोक्ताश्चतुर्दश ।
सुषुम्णेडा पिङ्गला च कुहूरथ सरस्वती ॥१५१॥

Now we explore the three main channels (veins/arteries) called 'nadi' in Sanskrit that are essential to understand the movement of life force in our physical bodies.

The 'Jivatma' embodied soul essentially uses the life force 'Prana' to ride up and down the central most important channel of energy transfer located in our spinal column. This is the 'Sushumna nadi' whereby the life force moves up to the 'Brahmarandhra' the cerebral aperture and back down to the base at 'Muladhara' chakra.   The operating deity of Sushumna is Lord Vishnu.   Sushumna is the channel that leads to the pathway to liberation 'Moksha'. Sushumna is aligned in the same center of the body with 'Brahmagranthi' and is flanked by 'Ida' nadi on the left and 'Pingala' nadi on right.

'Ida' is to the left of Sushumna and represents the feminine aspect of our existence. Vital breath that moves through Ida is that of Moon. 'Pingala' is to the right of Sushumna and represents the male aspect of our existence. Vital breath that moves through Pingala is that of the Sun. Ida extends from the Muladhara chakra to the tip of the left nostril and Pingala extends from the Muladhara chakra to the tip of the right nostril.

There are several more Nadis described in detail in the Ayurveda and yogic texts but for the scope of our analysis, the three described here are most important. (149-151)

The nadi's described here form the main foundation of understanding the flows of life force. In the practice of Yoga of many different types, what is controlled through yogic exercises is the flow of life force in varying degrees. The objective is to harness this energy so it can fuel our physical, mental, and spiritual development. To do this, we must intimately understand our psychosomatic body which we have discussed in detail in this section. Nadis are crucial to study as they represent the vast network of energy channels that makes each individual an integrated, conscious, and vital whole. ***Exploring them in further details is worthwhile as it would assist any practitioner of music to understand how the 'Naad' flows in their body***

The Sanskrit word nadi derives from the root word 'Naad', which means "flow," "motion," or "vibration." The word itself suggests the fundamental nature of a nadi: to flow like water, finding the path of least resistance and nourishing everything in its path. The nadis are our energetic irrigation system; in essence, they keep us alive. The three primary nadis that allow us to experience and control the flow of vital life force are the ones we described above, Sushumna, Ida, and Pingala.

The Sushumna (most gracious) nadi is the body's great river, running from the base of the spine to the crown of the head, passing through each of the seven chakras in its course. It is the channel through which kundalini Shakti (the latent serpent power) —and the higher spiritual consciousness it can fuel—rises up from its origin at the 'Muladhara' (root) chakra to its true home at the 'Brahmarandhra' at the crown of the head. In subtle body terms, the Sushumna nadi is the path to enlightenment.

The Ida (moon/Ganges/comfort/white) and Pingala (sun/Yamuna/tawny/orange) nadis spiral around the Sushumna nadi in a double helix format, crossing each other at every chakra. Caduceus, the symbol of modern medicine is an approximate visualization of the relationships among the Ida, Pingala, and Sushumna nadis. Eventually, all three meet at the 'Aagya' (command) chakra, between the eyebrows.

Just as Moon and Sun define the standards for our cycle of time so does the 'Prana' (life force) flowing through Ida and Pingala define the standards of 'Kaal' (time) of our life span. Sushumna 'destroys' time for when Kundalini flows through it from its source to the 'Brahmarandhra' the creative source of life of the embodied soul and the Supreme Brahman unites destroying any sense of duality. In that union of liberation, there is 'Advait', which is defined as non-duality between the creator and the creation.

The Ida nadi begins and ends on the left side of Sushumna. Ida is regarded as the lunar nadi, cool and nurturing by nature, and is said to control all mental processes and the more feminine aspects of our personality. The flow of energy in Ida is compared to the flow of river Ganges. The color white is used to represent the subtle vibrational quality of Ida. Pingala, the solar nadi, begins and ends to the right of Sushumna. It is warm and stimulating by nature, controls all vital somatic processes, and oversees the more masculine aspects of our personality. The vibrational quality of Pingala is represented by the color red. Flow of energy in Pingala is compared to the flow of river Yamuna.

The interaction between Ida and Pingala corresponds to the internal dance between intuition and rationality, consciousness and vital power, and the right and left brain hemispheres. Pingala is the extroverted, solar nadi, and corresponds to the left hemisphere. Ida is the introverted, lunar nadi, and refers to the right hemisphere of the brain. Ida Nadi controls all the mental processes while Pingala Nadi controls all the vital processes.

In everyday life, one of these nadis is always dominant. Although this dominance alternates throughout the day, one nadi tends to be ascendant more often and for longer periods than the other. This results in personality, behavior, and health issues that can be called Ida-like or Pingala-like. Creating Equilibrium between the Ida and Pingala is essential to mastering any form of musical or other art.

13 Music in Yogic Psychosomatic science

11. गीतादिसिद्धौ चक्राणं साधकत्वं बाधकत्वञ्च

अनाहतवले पूर्वेऽष्टमे चंकादशे तथा ।
द्वादशे च स्थितो जीवो गीतादेः सिद्धिमृच्छति ॥१४०॥

चतुर्यषष्ठदशमैर्दलैर्गीतावि नश्यति ।
विशुद्धेरष्टमादीनि दलान्यष्टौ श्रितानि तु ॥१४१॥

दद्युर्गीतादिसंसिद्धिं षोडशं तद्विनाशकम् ।
दशमैकादशे पत्रे ललनायां तु सिद्धिदे ॥१४२॥

नाशकं प्रथमं तुर्यं पञ्चमं च दलं विदुः ।
ब्रह्मरन्ध्रस्थितो जीवः सुधया संप्लुतो यथा ॥१४३॥

तुष्टो गीतादिकार्याणि सप्रकर्षाणि साधयेत् ।
एषां शेषेषु पत्रेषु चक्रेष्वन्येषु च स्थितः ॥१४४॥

जीवो गीतार्बिसंसिद्धिं न कदाचिदवाप्नुयात् ।

Having studied all the Psychosomatic and spiritual centers of our manifest body summary is provided of specific points therein relevant to musical arts. Focusing on the 1, 8, 11 and 12th petals of the Anahat chakra (cycle of unmanifest) empowers the embodied soul with full mastery over the music. Concentrating on 4, 6 and 10th petals must be avoided as they destroy musical ability. (140-141)

In 'Vishuddhi' chakra (cycle of purity) one must focus on petals 8 to 15 as they correspond to the seven swar and their tones. Whereas 16th petal must be avoided as it destroys musical ability. (141-142)

In the 'Kaal' or 'Lalna' chakra, the 10th and 11th petals enhance musical ability while 1, 4 and 5 are detrimental. (142-143)

Upon meditating and intimately knowing the 'Brahmarandhra' where thousand-petalled 'Sahastrapatra' resides the musical sadhak (practitioner) will find immense fulfillment and would progress in one's musical journey with ease and excellence. Emphasis on any other parts of the psychosomatic centers in terms of concentration would yield no musical fruits and might actually prove to be detrimental as stated in some specific cases here. (143-145)

We have covered in depth now the complete science of our psychosomatic composition and evolution. This knowledge is essential not just for those striving to practice music but for any curious seeker who is striving to understand one's true nature of evolution and the paths of progress in life. Audible 'Naad' is essentially nothing else but the Supreme Brahman manifested sensory audible musical form. In order to practice music in its fullness, one must understand the chakras mentioned above and their points of emphasis in our journey towards creating music that 'unites' us once again back to the Supreme Brahman.

At time of our birth through the detailed process discussed earlier our past actions and linkages to music manifest themselves in the present physical form our bodies and those who want to enhance and delve deeper into those past musical roots and continue this journey in this life and beyond, must study these psychosomatic linkages of points of petals to meditate upon.

Through either pure grace of the creator or spiritual, devotional and yogic deliberate efforts one achieves clarity of mind and spirit and reaches the highest point of 'Brahmarandhra' where one unites with the Supreme Creator'. At this point, there is nothing more to achieve as oneness with 'Brahman' frees up all other futile efforts of physical existence. At this point, it is the most opportune time for musical progress to create and offer the sweetest of music to the creator himself. Thus the musical journey and progress becomes effortless and spontaneous for the embodied soul who reaches 'Brahmarandhra'.

Physiological Sensory Motor tract and Chakra

The three main principles behind the subtle body and chakra systems are derived from the Indian Tantric Yoga system.

1. The mystical sounds of the Sanskrit alphabet are distributed across the 'petals' of all the chakras in the system,

2. Each chakra is associated with a specific Great Element (Earth, Water, Fire, Wind, and Space) and

3. Each chakra is associated with a specific Hindu deity or deities.

This is based on the objective of 'placement' in Tantra which is to visualize a specific sound tone syllable in a specific location in a specific chakra in your energy body while silently intoning its sound. This produces the needed effects of results as desired by the practitioner.

This practice is embedded in 'Naad' specific context in which the sounds of the Sanskrit language are seen as uniquely powerful vibrations that can form an effective part of an esoteric practice that brings about spiritual liberation or worldly benefits through spiritual energies.

Invoking the image and energy of a specific deity into a specific chakra is also particular to the Indian Vedic system.

The Aagya chakra and Manas chakra together form what we know in modern biological science as our sensory-motor tract. These two chakras and their interaction are directly related to our physical body and its functioning of the limbs that are instrumental in playing any of our musical instruments. Aagya chakra is the 'command center' that is connected to our motor nerves (aagyavaha nadis). On receiving the commands from the 'Aagya' chakra the motor nerves communicate their motor impulses to the limbs on our periphery for the execution of corresponding actions via the 'Manas' chakra.

'Manas' chakra is in turn connected to the corresponding sensory nerves coming directly from our periphery of limbs (of the five perceptible senses we have discussed earlier). These sensory nerves are; Olfactory sensory nerves (Gandhavah nadi), Optical sensory nerves (Rupavah nadi), Auditory sensory nerves (Shabdavaha nadi), Gastric sensory nerves (Rasovaha nadis), Tactile sensory nerves (Sparshvaha nadis). These nadis come from the periphery and are connected to 'Manas' chakra which is in turn situated at the bottom of our cerebrum in the sensory tract. Sixth sensory perception is that of intellect of the mind and is connected to 'Manas' chakra through sensory nerves directly linked to our brain (manovaha nadis) and these carry signals related our perceptions of dreams and other relevant states.

These facts help us in understanding the unique interrelation between the science of melody and rhythm and the yogic science of chakras and subtle body. Approaching music from these dimensions opens up new knowledge that is often not disclosed in texts.

Yoga and Esoteric Musical roots

We have described very useful details of our physiological and psychosomatic relations of Yoga, our evolution in the form of a human body and its significance for the musical science so far. The question that might arise in someone who has not studied these fields is, "Are these yogic and biological facts even essential to know if our objective is simply to produce a good melody or good rhythm in music?" To demonstrate why this knowledge is essential for a 'complete' seeker, let us consider a practical example that the Author has discovered by his research and study of these yogic, philosophical and musical schools of thought.

This knowledge is not documented in Sangeet Ratnakar but is presented here simply as an example of how valuable a holistic study of musical evolution is for a true seeker's progress. Being a practitioner of the Rhythmic as well as Melodic systems of music, the present author plays one of the oldest rhythmic instrument called Pakhawaj (known by its ancient name in Sanskrit as Mridang). The contemporary rhythmic instrument of choice in Indian classical Khayal style of singing is Tabla. Here our objective is not to discuss rhythm and its intricacies, but rather the esoteric connections between Yoga and Music leveraging the example of these Rhythmic instruments.

Teachers of Mridang and the musical texts tell us that the parent or root rhythmic cycle of Mridang is 12 beats and is known as Chautaal. Similarly, the teachers of Tabla and modern musical texts tell you that the root parent rhythmic cycle of Tabla instrument is 16 beats known as Teentaal or Tritaal. It is true that the beginner who learns these instruments must focus on these cycles as they are the 'master' rhythmic cycles of these instruments. But nowhere are the reasons documented as to 'WHY' this is the case. Many teachers also might not be aware of the root genesis of these deeper meanings of our Music.

Having discovered the underlying purpose of this and having discussed this with one of the most respected musical teachers in modern times, Padma Shri Dr. Rajeshwar Acharya, the Author presents the esoteric reason for the 12 beat and 16 beat parent taals of both Mridang and Tabla.

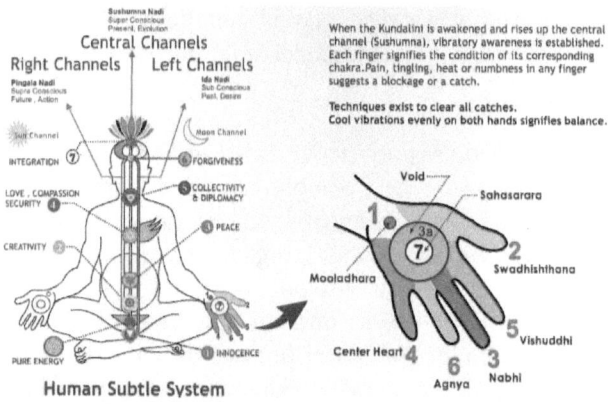

The logical and spiritual reason for the 12 and 16 beat root cycles for Mridang and Tabla respectively lies in our Tantric foundations of the yoga system. We have discussed yogic chakras and the nadis earlier. The above diagram summarizes the physiological interrelationship between these chakras and our hands. Now as one studies this closely the reasons for the parent rhythmic cycles of Mridang and Tabla slowly reveal to us.

Mridang – The main syllable sound of this Rhythmic musical instrument is 'Ta' and all other sounds of

Mridang are said to be relative to this 'Ta'. The part of our physiological body that is responsible to produce this on the instrument is the lower portion of our inside palm signified by color pink and chakra number 4 in the picture above.

This indicates that the main aspect of playing Mridang is leveraging this portion of our hand primarily and this happens to be connected to our Heart chakra called the Anahat chakra. This chakra has 12 petals and since the part of our hand that directly plays Mridang 'Ta' is connected to Anahat chakra, the 12 beat cycle naturally becomes the 'root' cycle of this instrument. *Thus Chautaal (12 beats) becomes the parent Taal of Mridang. This Taal is directly connected to our Heart and our heartbeats which have natural bpm measure of 72. (12 x6). It is for this reason that most vocalists when they practice Dhrupad form of singing in which Pakhawaj is the primary rhythmic instrument often first learn the singing in 12 beat cycles and then graduate to singing in other beat cycles.*

Tabla – The main syllable sound of this Rhythmic musical instrument is 'Na' and all other sounds of Tabla are said to be relative to this 'Na'. The part of our physiological body that is responsible to produce this on the instrument is the index finger (known in Sanskrit as 'Tarjini') denoted by the color blue and chakra number 5 in the picture above.

So, the leading physiological part that is responsible for playing this instrument is our index finger and this is connected to our Vishuddhi chakra or the Throat chakra. This chakra has 16 petals in it. Hence the index finger which is part of our hand that is controlled by this chakra in the subtle body leads the playing of Tabla. So to master Tabla the 16 beat cycles is the natural cycle

171

and this is in Tritaal or TeenTaal. *Thus Tritaal (16 beats) becomes the parent Taal of Tabla. This Taal is directly connected to our Throat and our vocal singing. It is for this reason that most vocalists when they practice Khayal singing in which Tabla is the primary rhythmic instrument often first learn the singing in 16 beat cycles and then graduate to singing in other beat cycles.*

The author arrived at the above crucial knowledge by the process of 'Self Realization'. It is not documented in ancient musical texts nor in the Tantric texts, but by striving to explore the principles of interconnectedness amongst the mind, body, music, and spirit, the Author derives this factual assessment.

Now another question arises that if such esoteric connections exist in rhythmic instruments do the other melodic instruments have these types of esoteric 'root' cycles. The answer is that these root 'Taal' idea applies only to the rhythmic cycles. When one considers the melodic instrument, the cycles can't be bound into only one root 'Raga' or melodic structure based on a given melodic instrument and its physiological method of playing. That would be against the principles of creativity of Melody.

However, in the case of Rhythm, the objective of 'Laya' is to put a boundary on the Melody so that the musical cycles become resonant, finite and enjoyable in an auditory sense by the harmony between melody and rhythm. This 'Laya' through Rhythm, can be applied through numeric counts whether it's in multiple of 12 or 16 and so on. Using these root cycles once mastered, a proficient musician can create many Laya cycles of differing numerical needs. For example, Jhaptaal is a 10 beat cycle but even using a 12 beat cycle of Chautaal one can arrive at the 12x5 = 60 (six Jhaptaal cycles equivalent rhythmic structure). As such for the Mridang and Tabla to have these parent 'root' Taal facilitates one's deeper understanding of the instrument and its tonality to master all other aspects of Laya. Melodic instruments depend upon the Laya foundation provided to them by this Mridang and Tabla through their Taals. A student of rhythm who understands these concepts and masters first the natural 12 and 16 beat cycles can achieve any future Laya numerical cycles as needed as described in Jhaptaal example above. But first, the root cycle must be understood and meditated upon.

This illustrates to us that approaching music holistically by understanding its spiritual and physiological relationships is tremendously valuable for the real seeker. Every detail in Indian music has a purpose and in this case, the answer is not found in musical text but rather by expanding our field of knowledge to the yogic texts and then connecting our musical knowledge to it. This is the process of self-discovery and self-realization in Music. Many aspects can't be learned from textbooks alone or from teachers alone, one must make genuine efforts to understand the roots of the knowledge and slowly the esoteric aspects start revealing themselves based on readiness and the objectives of the seeker.

Land Art depicting Chakras, Nadis and Kundalini

Nara, Japan

Purpose of music

ॐ उपसहार:

नादस्य भुक्तिमुक्तिसाधकत्वम्

एवंविधे तु देहे ऽस्मिन्मलसंचयसंवृते ॥१६३॥
प्रसाधयन्ति धीमन्तो भुक्ति मुक्तिमुपायतः ।
तत्र स्यात्सगुणाद्ध्यानाद्भुक्तिर्मुक्तिस्तु निर्गुणात् ॥१६४॥
ध्यानमेकाग्रचित्तैकसाध्यं न सुकरं नृणाम् ।
तस्मादत्र सुखोपग्यं श्रीमन्नादमनाहतम् ॥१६५॥
गुरूपदिष्टमार्गेण मुनयः समुपासते ।
क्षो ऽपि रक्तिविहीनत्वान्न मनोरञ्जको नृणाम् ॥१६६॥
तस्मादाहतनादस्य श्रुत्यादिद्वारतो ऽखिलम् ।
गेयं वितन्वतो लोकरञ्जनं भवभञ्जनम् ॥१६७॥
उत्पत्तिमभिधास्यामस्तथा श्रुत्याविहेतुताम् ।

While music has material value for enjoyment in this world "Bhukti", it is another usage is for spiritual salvation "Mukti". While it is easier for a common person to enjoy worldly music, it is very difficult for a common person to attain salvation using intense concentration needed to focus on formless Brahman with music as a medium.

Truly intelligent beings know even beyond Mukti and Bhukti, the most worthwhile purpose of music is for "Bhakti" which is the loving devotion of the Supreme Brahman in its many manifest and unmanifest forms. This devotional mood of the music is the easiest path of Enlightenment and union with the Creator by means of music.

Our physically manifested body has many impurities and imperfections and yet it is not to be discarded. Intelligent beings leverage our human form with proper care for the musical journey towards an objective of attaining union with the Supreme Creator. This is the secret. This is the essence of timeless music. (163-168)

Music is indeed a universal language. It can be used as a medium of materially aesthetic experiences as well as spiritually enlightening experiences. Through the ages across the world, music has been used as a powerful medium not just as a form of expressing human emotions but more universally as an instrument for spiritual and religious awakening. The Indian Vedic system and its scientific and spiritual principles place music in a prominent role in the pursuit of enlightenment especially in the present age known as 'Kaliyuga'. The most worthwhile endeavor for human existence is to identify and awaken the unmanifest "Anahat" Naad within one's body.

'Anahat' Naad is the Supreme Brahman in its subtle formless state and meditating upon it and contemplating it is extremely tedious. This is not easily achievable in the present age for an average person. However, the 'Ahata' naad is the one with 'form' it is audible and it can be experienced and its form contemplated upon and meditated upon. It is for this reason that the ancients on India created a systematic science of pursuing music through the forms of manifested Ahata Naad. This pursuit of the manifested Naad and intricately connecting it to spiritual ideals and objectives has been achieved by the 'Bhakti' (devotional) elements of the Indian scriptures. This is the easy path for an average person to embark upon and achieve the spiritual objectives thus.

Hence the purpose of our music (Science of our Melody and Rhythm) is to provide for us a very convenient and useful means to achieve worldly experience and pleasure through our senses (primarily led by our auditory sense objects) as well as achieve salvation and enlightenment by attaining oneness with the Supreme Brahman (primarily led by its Naad Brahman format). music in this regard has proven to be of immense utility universally as well as considered ideal for its attractiveness as a pleasurable and positive pursuit. It is in this regards very essential for a practitioner of music to study the Science of its Senses and Sensibilities from a logical as well as emotional perspective.

Swar Shastra Acharyas

So far, we have now understood music and its scientific, spiritual, philosophical and psychosomatic perspectives. This prepares us now to dive deeper into the practical and logical aspects of "Science of Melody" Swara Shashtra in ensuing chapters. What you will see is that our Indian sages profoundly created the systems of Swara Shastra based on logical and scientific techniques while simultaneously enhancing the Sensibilities (Emotional "Bhavatmak") aspects of the Art. *A seeker of true Melody must understand that this is a journey of multiple lifetimes and the ocean is too deep for one lifetime.*

Let us first remember the founding fathers of this system. In the Indian spiritual sense, music has several ancient originating main schools of approach dating back to the Creator. *The original "Acharyas" (Gurus) of Indian musical knowledge are considered the founding creators of the Science of Melody* in Indian musical context. They are named as follows;

1. Shiv (Sadashiv)
2. Vishnu (Hari manifested as Shri Ram and Shri Krishna)
3. Brahma
4. Sage Bharat Rishi
5. Sage Kashyap Muni

Lord Shri Ram's Eternal Devotee Pavanputra Shri Hanumanji

6. Pavanputra Shri Hanumanji
7. Sage Angad
8. Sage Narad Muni
9. Gandharva Tumbru
10. Goddess Durga
11. Goddess Parvati

178

12. Sage Matang
13. Sage Yask
14. Sage Shardul
15. Sage Kohal

A musician who understands the above scientific and spiritual dimensions described herein will have a better chance of understanding the pursuit of Naad Brahman. If one practices music with the sincere humility and understanding of the ideas described here, such a person would have infinite possibilities and wonderful opportunities of progressing on the path of Naad Yoga. What is required is an intense commitment and spiritual thirst to go to the deepest depths of the infinitely deep "Ocean of Laya and Swara". In that journey itself, one finds the point of unification with each note and each beat created by the Supreme Brahman.

Section 2 - Naad, Shruti, Swara

Naad (Nada)

Nada Brahma and its Worship

अथ तृतीयं नादस्थानश्रुतिस्वरजातिकुल-
दैवतर्षिच्छन्दोरसप्रकरणम्

क. नादः

(i) नादब्रह्म, तस्योपासना च

चैतन्यं सर्वभूतानां विवृत्तं जगदात्मना ।
नादब्रह्म   तदानन्दमद्वितीयमुपास्महे ॥१॥
नादोपासनया देवा ब्रह्मविष्णुमहेश्वराः ।
भवन्त्युपासिता नूनं यस्मादेते तदात्मकाः ॥२॥

Let us worship the "Nada-Brahman" which is the incomparable bliss form of the creator himself and is omnipresent in all living beings, all intelligence and the whole being of the universe. It is important to note here that while the words Nada and Brahman are separate, in the context of the term "Nada-Brahman" they are used as a compound word meaning Nada is nothing but a manifestation of the Brahman itself. (1)

Through this worship of the Nada, we are worshiping none other than the trinity of Brahma, Vishnu, and

Mahesh as they are related in the same manner, as a ray of Sun is the same as Sun. They are the same. Nada Brahman is the all-pervading undifferentiated manifestation of the Creator. (2)

It is important to note here that when we use the term Nada in the macro context as above it encompasses both "Anahata" Nada as well as "Aahat" Nada, which is audible. In general, Dr. Shringy aptly describes sense Nada in his English transliteration of Sangeet Ratnakar as "Consciousness of Sound". The author also subscribes to this appropriate definition. However, pursuant to this section as we describe other characteristics of Nada and define other aspects of Science of Melody for all practical purposes unless stated otherwise, the term Nada refers to the "Ahata" variety of Nada that is the audible "musical Sound"

Inner Psychosomatic Manifestation of Nada

This process has been described in detail in our earlier volume 1 of this Nada Yoga Series "Science of Rhythm". However, for cross reference, we summarize it in this volume as well. Essentially the process for ALL music generation in our physically manifested body is from the same original flow of steps as summarized here by Sharangdev.

(ii) देहे ध्वनेराविर्भावः

आत्मा विवक्षमाणोऽयं मनः प्रेरयते, मनः ।
देहस्थं वह्निमाहन्ति स प्रेरयति मारुतम् ॥३॥

ब्रह्मग्रन्थिस्थितः सोऽथ क्रमादूर्ध्वपथे चरन् ।
नाभिहृत्कण्ठमूर्धास्येष्वाविर्भावयति ध्वनिम् ॥४॥

181

Up to this point in our book, we cover the Nada in its entirety emphasizing its source of origins and creation of the universe and the links to Anahata Nada. From this point onwards in verse 3 and 4 now we will start the journey of exploring all facets of Nada that is physically manifested and is enjoyed by living beings. This is the Aahat Nada as we have described earlier and now we summarize the process of its creation from the point of view of a human physical body.

The individual being "Atma" first desires to speak and as a result, instructs the mind as such. Mind, in turn, activates the energy power center in the human physical form stimulating the vital forces of the body to initiate voice generation needed for music creation. (3)

This vital force is situated in an unmanifested form in our physical body below the navel region. When activated as per Atma's wishes above, the vital force gradually travels upwards from Navel onwards in our body and gradually manifests itself as "Aahat Nada" which produces our speech essential for any creative music. The chain of travel of this Nada upwards is through the five stages of physical body navel, heart, throat, cerebrum and finally mouth cavity. (4)

Fivefold Nada

(iii) पञ्चविधो नादः

नादोऽतिसूक्ष्मः सूक्ष्मश्च पुष्टोऽपुष्टश्च कृत्रिमः ।
इति पञ्चाभिधा धत्ते पञ्चस्थानस्थितः क्रमात् ॥५॥

As the Nada evolves upstream from Navel, it passes through five stages of gradual manifestation at each of the five physical locations in our body.

I.   Navel – Ati Sukshma Nada (extremely subtle) – This is the origin of Nada manifestation and hence it is extremely small as it is being manifested in the navel.

II. Heart – Sukshma Nada (subtle) – As it travels upstream Nada becomes slightly larger in its manifestation in our hearts.

III. Throat – Pushta Nada (Richer or Louder nada) – In our Throats (vocal chords) the Nada now takes a larger and louder form.

IV. Cerebrum – Apushta Nada (no so loud or faint Nada) – As the cerebrum is the highest and farthest point in our body where Nada travels it is at that point farthest from its origins in the Navel region hence it is fainter naturally at that physical location and is also intuitively called mental speech which where Nada still retains its purity and full character as envisioned by our Atma.

V.   Mouth Cavity – Krutrim (Modified) Nada – Finally from the brain Nada travels to our mouth and here the true nature of Nada is modified and hence it is called Modified in nature. A question may arise as to why the speech produced from our mouth is called artificial from the perspective of Nada manifestation.

The reason is evident when one analyses the fact that after the sound is produced in our throat and continues upwards in the brain, the mouth has various aspects that modify the sound such as teeth, tongue, lips, the physical health of the body producing the sound from that mouth, etc. All these factors "modify' the natural intended pure form of Nada and hence what comes out of our mouth is essentially Nada that has lost some of its naturally intended format. (5)

There are some minor differences in the order of the first two of these as to which navel or heart is the Sukshma or Ati Sukshma nada. Sharangdev and Matanga both differ slightly here but as Sharangdev was also a Doctor of Ayurveda, the present author feels that his definition of the order given here is a good point of reference. Also, many experts might suggest that the Nada produced from Mouth is Artificial, however, the present author feels that the term Modified defines it correctly and avoids any misconception with the term "artificial" as this Nada is also real, it's just that it is modified from its true form due to the factors described above.

Derivation of Nada

(iv) नादशब्दस्य निरुक्तिः

नकारं प्राणनामानं दकारमनलं विदुः ।
जातः प्राणाग्निसंयोगात्तेन नादोऽभिधीयते ॥६॥

T

his Nada comprises of two words literally in its sound. "Na" and "Da". (It is important to observe that in Sanskrit the derivations of words can be analyzed in two methods one that is grammatical "vyutpatti" and one, which is Semantic "nirukti") Semantically the syllable "Na" or the letter "n" in the Hindu Tantra tradition represents "Prana" or vital force. The syllable "Da" or letter "d" refers to "Agni" which is fire. Hence, this yin and yang "union" of Vital force and Fire in our body produces the Nada, which is essential, what we know as "Consciousness of our Sound". (6)

Threefold Nada Musical Practice

(v) गीतव्यवहारे त्रिधा नाद:

व्यवहारे त्वसौ त्रेधा हृदि मन्द्रोऽभिधीयते ।
कण्ठे मध्यो मूर्ध्नि तारो द्विगुणश्रोत्तरोत्तर: ।।७।।

In actual practice, this Nada takes practically threefold form. Moreover, each of these three forms takes on a double pitch ratio from the previous and has a relation 1:2:4 in the matter of pitch intensity.

a) Heart – Mandra Naad (1 intensity) - Lower pitched or lower octave

b) Throat – Madhya Naad (2 Intensity) – Medium or normal octave.

c) Head – Tara Naad (4 Intensity) – Higher pitched or Higher octave

The earlier fivefold classification is very useful for understanding the precise internal evolution of speech and Nada production. However, for singing and music which is heard externally, this threefold practical grouping of Nada manifestation is more relevant. The intensity or double ratio relationship can be further understood as the effort needed to produce notes of a given register. So the effort needed to produce notes of Madhya Octave register is double that of Mandra register. And Effort required to physically produce the sound of Taar (higher octave) is double that of the Madhya (middle register octave) (7)

Shruti

Shruti Definition

ख. श्रुतिः

(i) श्रुतिः, तत्सङ्ख्या च

तस्य द्वाविंशतिभेंदाः श्रवणाच्छ्रुतयो मताः ।
हृद्घूर्ध्वनाडीसंलग्ना नाङ्घो द्वाविंशतिर्मंताः ॥८॥
तिरश्च्यस्तासु तावत्यः श्रुतयो मारुताहतेः ।
उच्चोच्चतरतायुक्ताः प्रभवन्त्युत्तरोत्तरम् ॥९॥
एवं कण्ठे तथा शीर्षे श्रुतिर्द्वाविंशतिर्मंता ।

In the Nada Section, we defined threefold practical categories of Ahata Nada produced by Heart, Throat and our Brain. Now this Manifested sound is further crystallized into what we call Shrutis. The word Shruti in Sanskrit can be semantically transliterated as "Audible". So based on various degrees of Audibility Nada is further practically classified into 22 Shrutis. (8)

Physiologically in our body, there are two main arteries connected to our heart per the Indian Vedic medical science. "Ida" and "Pingala". These two are intersected with 22 Horizontal "Nadis" or arteries in turn and these 22 arteries produce the 22 Shruti successively with the force of the wind acting upon the sound as it travels upwards through the main arteries of the heart. (9)

We discussed Heart (Mandra Register – Lower Octave) above, but let us remember that the threefold Nada is produced in three registers of Heart, Throat, and Cerebrum. Hence the 22 Nadis or 22 Shruti are also similarly produced in other two registers of Throat (Madhya Register – Medium Octave) and Cerebrum (Taar Register Higher Octave) (10)

"Shruti" and their meaning is the second most fundamental concept in understanding Indian musical Science. First being the Nada itself. As such, Shruti must be explored in further depth. As it is understanding and mastery of these Shrutis that makes a performer's music that much more deep and richer as opposed to just mimicking the notes.

Here a key point must be made that as we explore Shrutis one cannot but avoid references to "Swara" even though the subject of "Swara" will get its own independent analysis in the immediate section pursuant to the Shruti. The reason why Shruti and Swara discussions go hand in hand and are often circularly pointing to each other is that one cannot exist without the other. Shrutis essentially when grouped together form the "Swara' and Swara alone has no meaning unless it's encompassing Shruti structure is understood. As such, the readers in this and the Swara sections are advised to revisit at comfort both sections as needed to solidify their understanding of both these subjects. Study and re-study them again for a complete holistic understanding.

The word "Shruti" as we have discussed is derived from the Sanskrit root "sru" which means to hear. Therefore, Shruti is the name given to the range of sound that is manifested and is possible to be heard by a listener. To define the Shruti precisely after many millennia of expert analysis and scholarly studies, Since the time of

Sharangdev's Sangeet Ratnakar the Indian musicologists have all reached consensus as the number of Shruti to be 22.

"Shruti is that audible sound which at the conceptual level is capable of being individually perceived, recognized and reproduced. At a Perceptual level, it is free from Resonance and is the first sound before other tonal characteristics are superimposed upon it." (From Dr. R.K. Shringy's commentary).

Conceptually Shrutis are now agreed upon as 22 and they occur in all the three registers lower, middle and higher registers at precise and definite positions in the scale. Perceptually, Shrutis are non-resonant primary manifestations of musical sound, *antecedent to the production of subsequent Swara (tones).* This is the pure technical definition of Shruti. In the English language, a synonym for Shruti is "microtone". It is these Shrutis and their reproduction by experts that help in defining some music to be sweeter than others do. The number of Shrutis in successive combinations allow us to perceive and produce a tone.

22 Shruti Groupings

Several ancient scriptures group Shrutis based on three places where the voice is produced as we discussed in Nada viz. Heart, Throat, and Cerebrum. This classification aligns naturally with three registers of notes related to these three places in our body.

Another grouping of Shrutis is based on the actual Fourfold condition of our body based on the predominance of "Vata" (Wind or Gas), "Pitta" (Bile) and "Kaf" (Phlegm or that which causes Cough). Permutations and combinations of these three naturally occurring bodily elements give rise to corresponding Shrutis. In other words, these three also are corresponding often to what we know as Timbre, Pitch and Volume. These three qualities of Sound in our Shruti are directly proportional to the balance/imbalance of the three bodily elements indicated. The above classifications help us in understanding the physiological theory of the Shruti groupings.

More practically speaking common groupings that matter to us are the ones based on the Pitch of the notes. When specifically grouped based on the pitch levels, these Shrutis then give birth to the "Swara" as we have mentioned earlier. Now let us understand these Groups. There are essentially three groupings of Shrutis comprising of corresponding "Swara"

i. 4 Shruti Interval Swaras – Shadaj (s), Madhyam (m) and Pancham (p)

ii. 3 Shruti Interval Swaras – Rishabh (r), and Dhaivat (d)

iii. 2 Shruti Interval Swaras – Gandhar (g) and Nishad (n)

Thus, the First group of Shrutis is composed of Total 12 Shruti microtones with three Swara of 4 microtones each.

Thus, the Second group of Shrutis is composed of a Total of 6 Shruti microtones with two Swara of 3 microtones each.

Thus, the Third group of Shrutis is composed of Total 4 Shruti microtones with two Swara of 2 microtones each.

Thus, we total 22 Shrutis covering all Seven Swara, as we know them. "Swara" we will define later as the sound that immediately follows the Shruti in order of its resonance in three groups of Shrutis respectively. In other words, starting from Shadaj (s) this Swara comprises the first 4 Shrutis, then Rishabh (r), comprises the next three Shrutis in the sequence of 22 Shrutis and so on.

Physiological Evolution

Many contemporary scholars of Sharangdev have understood the Nadi and Shruti relation as described in verse 8 above. The precise physiological connection of these sounds to our body was further clarified by "Sharadaranya" a contemporary of Sharangdev who takes the definition of Shruti and their physiological evolution and completes it with identifying even the Swara. The first question that arises is are there only 22 Shrutis and through proof from scientific experiments, it has been proven now that the number 22 of Shrutis is the widely acknowledged baseline. This fact is also equally aligned with the Physical condition of the Human Body as explored below.

Three main arteries pass through our Heart as per the ancient Indian system of Ayurveda medical science. They are Ida, Pingala and Sushumna. Ancient sages identified 22 Shrutis as the veins crossing these three main arteries going vertically upwards through the heart. Per Sharadaranya, he defines the 22 Shrutis as the 22 cross-referencing veins crossing the vertical main artery Sushumna passing through the heart. He further identifies seven locations in our body where the seven alphabetical "Varna" or pronunciations are produced. The table below will act as a future reference for meditating upon these connections.

Table of Shruti Swar Varnasthan

No.	Swara	Shrutis	Varnasthan (Physiological Location) of Pronounciation	Nadis attached to Shushumna
1	Shadaj	4	Throat	4
2	Rishabh	3	Root of Palate	3
3	Gandhar	2	Lips	2
4	Madhyam	4	Center of Cerebrum	4
5	Pancham	4	Teeth plus all of above	4
6	Dhaivat	3	Throat and Palate	3
7	Nishad	2	Throat and Lips	2

Thus, Sharadaranya not only identifies 22 Nadis related to the 22 Shrutis but also completes the picture by cross-referencing Seven Swara as well in our body and their places of pronunciation.

This Physiological connection is further developed and extended by understanding the relationship between the "Dhatus" seven fundamental "supportive tissues" composing our body and the Swara and Shruti related to them. Essentially, when Sound is produced through our physiological bodies, energy transformation is occurring that is converting internal energy into external audible energy.

This transformation occurs using two ingredients of Fire and Air. The Fire is provided by our body's natural "Ushma" internal bodily heat and acts upon the main life source which is the "Prana" which is also closely related to our breathing and the "Vayu" (Oxygen) being inhaled and exhaled by us. This heating of the "Prana" by our "Ushma" occurs in various places in our physical bodies as described in the table above. These places are composed of supportive body tissue matter and based on the structure of this tissue the sound energy produced therein rises upwards in the 22 "Dhamanis" (Arteries) or commonly known as Nadis also. The word Dhamani in Sanskrit signifies a pump pumping air out. Thus, the sound and music are produced by the internal energy of our "Ushma" heating our "Prana" in different places composed of different supportive tissues and then is pushed externally through the route of the 22 "Dhamani" (Arteries).

This is a complete physiological understanding of the deep Indian knowledge of Sounds being produced by our body.

Shruti Swar Dhatu Table

No.	Swara	Shrutis	Dhatu	Dhamanis	Substratum Location
1	Shadaj	4	Semen	4	Brahmagranthi Yogic Center of Body
2	Rishabh	3	Marrow	3	Navel
3	Gandhar	2	Bone	2	Heart
4	Madhyam	4	Fat	4	Throat
5	Pancham	4	Flesh	4	Root of Palate
6	Dhaivat	3	Blood	3	Cerebrum
7	Nishad	2	Skin	2	Sahastradhara Chakra
	Total	22		22	

The above Shruti microtones and their related Swara tones are thus connected to the resident "Ushma" or Fire in these seven types of tissues but do not have anything to do with tissue themselves. The key is the differing degrees of Fire residents in these Tissues. When this Fire is fed from the "Dhamani" the Arterial passages, then the tones are produced as a modification of the Ahata Nada we have described earlier and this Nada is primarily traversing through the main Nadi Sushumna.

Fourfold String movement

Thus far, we have seen that the 22 Shrutis are based on the 22 Nadis in our human body and this is not just based on Yogic and Tantric spiritual principles but is also measurable by the scientific and objective process to be described next. As we shall see the 22 Shrutis can be clearly be produced and reproduced as individual entities using Vinas in the measurable experiment of Fourfold string movements that follows.

A scientific process has to be a repeatable experiment that anyone can perform in the right conditions. While discussing purely technical details of producing and conducting the experiment would deviate us too much from the matter of the overall details of Science of Melody, we summarize the approach here for the curious seeker. This experiment is done using what is known as the Fourfold String movement on two Vinas. There are other methods of experimenting and proving the number of Shrutis is 22 as well and in the future, with newer scientific developments more methods will be produced.

However, the process summarized briefly is the ancient process as documented by the Indian musical sages. As we shall see the 22 Shrutis can be clearly be produced and reproduced as individual entities using Vinas in the measurable experiment of Fourfold string movements that follows.

(ii) चतुःसारणा

व्यक्तये कुर्महे तासां वीणाद्वन्द्वे निदर्शनम् ॥१०॥

द्वे वीण्ये सदृशौ कार्ये यथा नादः समो भवेत् ।
तयोर्द्वार्विंशतिस्तन्त्र्यः प्रत्येकं, तासु चाविमा ॥११॥

The Shrutis mentioned earlier are 22 and any music must leverage them to produce more enhanced emotions. However, some Shruti intervals are so minute that vocally without tremendous expertise it might not be possible to reproduce them. Hence, a point of reference is required to study, learn and understand the Shrutis in all the three registers. This is where the instrument Vina comes into tremendous usefulness since ancient times. In this Fourfold string movement, we now discuss the scientific process of analyzing the Shrutis using the mother of all plucked string instruments the Vina that is the most ancient of them all. Vina is also considered the main instrument of goddess Sarasvati in Hindu theology and she is the goddess of all knowledge and fine arts.

The process described in detail is in Sharangdev's SR and is pertaining specifically to Vina experimentation. Rather than describe the entire experiment here we summarize the findings here. This is but one tried and tested example from sages of the past to understand the Shrutis. There are perhaps better or newer processes using modern technology that one can leverage now or in the future to measure the same Shrutis. The key point is that each of the Shruti can be scientifically measured and identified.

For understanding and their clear production, two Vinas are to be used with 22 strings in each of the Vinas. The vinas must be manufactured to be identical in all aspects to sound exactly alike. The first string must be tuned to produce the lowest possible audible and musically delightful sound. (10, 11)

The Second string must be tuned to produce a bit higher sound than the first one maintaining continuity of progression between first and second string and not allowing any other audible sound. Thus these strings fixed one below the other are of successively higher pitch and the sound produced as such is the 22 "Shrutis".

The result of anyone wanting to leverage two Vinas to perform the precise experiment of the Four-Fold String movement is that the number of Shrutis is exactly measured and verified to be 22 in nature.

Swara

7 Swara (Tones) Definition

ग. स्वरः

(i) सप्त स्वराः

श्रुतिभ्यः स्युः स्वराः षड्जर्षभगान्धारमध्यमाः ।
पञ्चमो धैवतश्चाथ निषाद इति सप्त ते ॥२३॥
तेषां संज्ञाः सरिगमपधनीत्यपरा मताः ।

From out of the Shrutis arise the Swaras. These are seven as known by their names, Shadaj, Rishabh, Gandhar, Madhyam, Pancham, Dhaivat, and Nishad. Their other accepted naming convention is "Sa-Ri-Ga-Ma-Pa-Dha-Ni" (23)

Abhinav Gupt in his treatise on Indian music defines Swara as "That sound which is produced pursuant to and by the influence of the initial sound arising as a result of striking a (vibrating string) on a Vina at a Shruti-Sthan (position of origination of a Shruti). This sound is essentially a resonating sweet and creamy (mellifluous)." This is the technical definition of Swara. Note that Shruti in its first instance of the strike is non-resonating but it produces vibrations in the string which then gives "birth" to the Swara noted above. The Shruti itself is nonresonant as we have discussed earlier, but it gives birth to Swaras which have the Tonal Triad characteristics of Uddat, Anuddat, and Svarit.

There are Seven Swara and it would be pertinent here to define the names and qualities of each swara described above. For this, the best reference available is from Matanga's Brhdessi treatise.

Shadaj

It is so-called, as it is the father of all the other six Swaras that come after it. In other words, Shadaj contains the sounds of all the other notes. All six other notes reside in Shadaj and hence the name. There is also a physiological meaning behind Shadaj and its name. Six organs of our body are used to produce the Shadaj sound and hence its name i.e. Nostrils, Throat, Palate, Brest, Tongue and the Teeth.

Rishabh

It is named as such, as it means, "Quickly appeals to our Heart". In nature an example is given of a cowherd whereby a Bull in the herd is distinctively strong, so is Rishabh amongst the other notes distinct due to its strength. Rishabh, of course, means a Bull as well in Sanskrit and hence the sound that is as strong as a bull is Rishabh.

Gandhar

Gandhar is named as such because of its Sanskrit root meaning "It holds musical speech". Secondly, it is said to be the sound that pleases the Gandharva (Semi divine celestial beings who are masters of music)

Madhyam

Madhyam in Sanskrit literally means in the middle literally. In addition, this note is in the middle with three on each side of it in the Seven note register and hence the name.

Pancham

This is the fifth note from the Fundamental note Shadaj. Hence, the name Pancham derived from number five. Pancham is also that which helps in measuring the relative extent of other note's sounds.

Dhaivat

It is named as such, as it is discerned by minds that are very sensitive in their nature. Shrutis of the posterior tonality (Uttaranga) in the register bring about this note.

Nishad

It is named as such, as it literally means "NI" and "Sad" to come to a rest. Notes of the Indian melodic scale register "Saptak" (heptad) end on the Ni as the last note and the upper register Shadaj restarts the second register. Thus, Nishad is named due to its position in the Saptak.

Shruti Swara relationship

(ii) स्वरलक्षणम्

श्रुत्यनन्तरभावी यः स्निग्धोऽनुरणनात्मकः ॥२४॥

स्वतो रञ्जयति श्रोतृचित्तं स स्वर उच्यते ।

Immediately pursuant to the Shruti, production, creamy in nature, resonating and delightful independently on its own for the minds of the listener is the "Swar". (24)

Here we reinforce the definition of Swara and build upon it further to solidify their character in our minds. This definition again also illustrates the difference between Shruti and Swara. As stated earlier both these concepts of Shruti and Swara are two sides of the same coin and hence every time ones speak of either, the other aspect must be cross-referenced. Shrutis essentially are the mothers of Swara.

When a string of Vina is plucked the first sound produced is the Shruti and this is a non-resonating sound. The very next sound that is then produced due to vibrating resonance of the first Shruti and subsequent resonance is the Swara. Resonance hence is the essential quality of Swara while Shruti is free from it. The other two most important qualities needed to define a Swara are Softness "Mriduta" and self-sustained delightfulness i.e. without needing any other accompanying help for the listener.

So one can surmise now that Swara essentially is Shruti that is further "developed" - If Shruti is the first sound generated by plucking the chord of a Vina then Swara is the further development of that sound whereby it gathers resonance and develops further enhanced sonorous qualities. This further enhanced "post-sound" is also known as "Anurana" in the SR verse 24 above.

Swara and Shruti allocation

(iii) श्रुतीनां स्वरकारणत्वम्

ननु श्रुतिश्चातुर्थ्याविरस्त्वेवं स्वरकारणम् ॥२५॥
श्र्यादीनां तत्र पूर्वासां श्रुतीनां हेतुता कथम् ।
भूमस्तुर्यातृतीयाऽऽदिः श्रुतिः पूर्वाऽभिकाङ्क्षया ॥२६॥
निर्धार्यंतेऽतः श्रुतयः पूर्वा अप्यत्र हेतवः ।

We have established the Shruti and Swara and their relationship and now the following important question may arise. Since Shrutis have 3 groupings as we have seen if the 4th (3rd and 2nd) Shruti in the respective grouping produces Swara then what is the purpose of other remaining intermediary Shrutis? (25)

Now lets us clarify this question as such. Such Third, Fourth or other intermediary Shrutis are determined ONLY in relation to their preceding Shrutis, hence they can also constitute the Swara. In other words, the 22 Shrutis are a cyclical loop depending on where you start the relationship of 4th 3rd. etc. can shift and also produce Swara. (26)

There is no doubt whatsoever that Shruti as a whole brings about and gives birth to Swara. We also know that Shrutis are grouped into three groupings of 4 Interval, 3 Interval and 2 Interval Shrutis and Swars are made from these groups. As such Swaras of 4 Shruti are produced by 4th Shruti etc. However, the Swara of 4 Shruti means that it has three other Shrutis preceding it.

Thus, no matter which point in the 22 Shruti continuum you start as long as the point has necessary 4, 3 or 2 Shrutis preceding it then that Shruti can be said to give birth to that Swara.

The author suggests here that readers cross-reference the Shruti Swar Table below to understand this relativity of Shruti and Swara. The Swara here in the table below can be movable vertically up and down the Shruti chain as long as the preceding Shruti conditions are met.

3 Shruti Groups	Shruti Microtone Number	Swara Note	Final Shruti of Note in Scale
4 Shruti Interval	1	Shadaj	
4 Shruti Interval	2	Shadaj	
4 Shruti Interval	3	Shadaj	
4 Shruti Interval	4	**Shadaj**	**4**
3 Shruti Interval	5	Rishabh	
3 Shruti Interval	6	Rishabh	
3 Shruti Interval	7	**Rishabh**	**7**
2 Shruti Interval	8	Gandhar	
2 Shruti Interval	9	**Gandhar**	**9**
4 Shruti Interval	10	Madhyam	
4 Shruti Interval	11	Madhyam	
4 Shruti Interval	12	Madhyam	
4 Shruti Interval	13	**Madhyam**	**13**
4 Shruti Interval	14	Pancham	
4 Shruti Interval	15	Pancham	
4 Shruti Interval	16	Pancham	
4 Shruti Interval	17	**Pancham**	**17**
3 Shruti Interval	18	Dhaivat	
3 Shruti Interval	19	Dhaivat	
3 Shruti Interval	20	**Dhaivat**	**20**
2 Shruti Interval	21	Nishad	
2 Shruti Interval	22	**Nishad**	**22**

Shruti Classifications

(iv) पञ्चजातिषु सप्तस्वरेषु च सनाम श्रुति-विभाजनम्

दीप्ताऽऽयता च करुणा मृदुर्मध्येति जातयः ॥२७॥
श्रुतीनां पञ्च तासां च स्वरेष्वेवं व्यवस्थितिः ।
दीप्ताऽऽयता मृदुर्मध्या षड्जे स्यादृषभे पुनः ॥२८॥
संस्थिता करुणा मध्या मृदुर्गान्धारके पुनः ।
दीप्ताऽऽयते मध्यमे ते मृदुमध्ये च संस्थिते ॥२९॥
मृदुर्मध्याऽऽयताऽऽख्या च करुणा पञ्चमे स्थिता ।
करुणा चायता मध्या धंवते सप्तमे पुनः ॥३०॥
दीप्ता मध्येति तासां च जातीनां ब्रूमहे भिदाः ।
तीव्रा रौद्री वज्रिकाग्रेत्युक्ता दीप्ता चतुर्बिधा ॥३१॥
कुमुद्ध्यायता याऽस्याः क्रोधा चाथ प्रसारिणी ।
संदीपनी रोहिणी च भेदाः पञ्चेति कीर्तिताः ॥३२॥
दयावती तथाऽऽलापिन्यथ प्रोक्ता मदन्तिका ।
त्रयस्ते करुणाभेदा मृदोर्भेदचतुष्टयम् ॥३३॥

मन्दा च रतिका प्रीतिः क्षितिर्मध्या तु षड्बिधा ।
छन्दोवती रञ्जनी च मार्जनी रक्तिका तथा ॥३४॥
रम्या च क्षोभिणीत्यासामथ ब्रूमः स्वरस्थितिम् ।
तीव्राकुमुद्वतीमन्दाच्छन्दोवत्यस्तु षड्जगाः ॥३५॥
दयावती रञ्जनी च रक्तिका चर्षभे स्थिताः ।
रौद्री क्रोधा च गान्धारे वज्रिकाऽऽथ प्रसारिणी ॥३६॥
प्रीतिश्च मार्जनीत्येताः श्रुतयो मध्यमाश्रिताः ।
क्षिती रक्ता च संदोपन्यालापिन्यपि पञ्चमे ॥३७॥
मदन्ती रोहिणी रम्येत्पेतास्तिस्रस्तु धंवते ।
उग्रा च क्षोभिणीति द्वे निषादे बसतः श्रुती ॥३८॥

203

Now we define the FIVE classes of Shrutis. (27-38)

1. Dipta (Dazzling lit. illuminated)

2. Ayata (Vast lit. extended)

3. Mridu (Tender lit. soft)

4. Madhya (Moderate lit. medium or central)

5. Karuna (Compassion lit. sadder emotion)

The Above 5 are further sub-classified into names corresponding further to precise tonal emotional characteristics and their groupings are further defined as;

  i. Dipta - Fourfold (tivra, raudri, vajrika, ugra)

 ii. Ayata – Fivefold (kumudvati, krodha, prasarini, sandipani, rohini)

iii. Mridu– Fourfold (manda, ratika, priti, ksiti)

iv. Madhya – Six (chandovati, ranjani, marjani, raktika, ramya, kshobini)

 v. Karuna – Threefold (dayavati, alapini, madantika) (29-38)

Note: Each of the names above signifies a very specific "Bhavatmak" emotional tonal quality and their true nature and depth can only be experienced by meditative soul devoted to exploring the innermost nature of Nada Yoga. If in one lifetime, one can experience even one of these Shruti sub-classifications in their practice and performance then their efforts would deem to be worthwhile and on the right track. The above names are of usefulness to those who want to dive into innermost elements on the Indian Science of music.

The classification of Shrutis is based on the Indian science of Aesthetics ("RasaShashtra"). In RasaShashtra, all the human moods and emotions of life are depicted and their corresponding characteristics in art forms are defined. The Shrutis essentially are the most granular forms of depicting emotions of "Rasa" through Audible qualities. As such, this classification is of paramount importance for those interested in studying and practicing Indian music based on exploring and highlighting the "Bhava" (Emotional Quotient) aspects of the music.

The Five main categories above are grouped by their assignments to the seven Swara heptad first. Each Shruti first belongs to a common class that signifies their main overarching emotion. Then the sub-classification highlights further special traits and emotional qualities of the Shrutis. In addition, the number of special moods assigned to each of the Five Classes directly corresponds to the number of Swars that the class of Shruti is assigned to. E.g. Dipta is assigned to Four Swara and is seen to have four sub qualities defined. Karuna is assigned to Three Swara and is defined to have three sub-classifications and so on. This entire structure is clarified in the table that follows;

| | | Shruti Classification Table | | | |
		1	2	3	4	5
	Swara	*Dipta*	*Ayata*	*Mridu*	*Madhya*	*Karuna*
1	Shadaj	tivra (1)	kumudvati (2)	manda (3)	chandovati (4)	-
2	Rishabh	-	-	ratika(3)	ranjani (2)	dayavati (1)
3	Gandhar	raudri (1)	krodha (2)	-	-	-
4	Madhyam	vajrika (1)	prasarini (2)	priti (3)	marjani (4)	-
5	Pancham	-	sandipani (4)	ksiti (3)	raktika (2)	alapini (1)
6	Dhaivat	-	rohini (3)	-	ramya (2)	madantika (1)
7	Nishad	ugra (1)	-	-	kshobhini (2)	-

Here in this table, it is important to note that the five main Shruti "Rasa" classifications are also sub-classified into finer emotions. Each of these finer emotions is listed in the "ORDER" of the Shruti structure of that Swara. E.g. Shadaj is a 4 interval Shruti type Swara composed of 4 Shrutis and those are precisely defined in the order of those four Shrutis and their names with the number of the Shruti sequence in brackets. So First of the four Shrutis of Shadaj is Tivra (Dipta), 2nd is Kumudvati (Ayata), etc.

Take Pancham as an example and one can see the First order Shruti in Pancham is Alapini (Karuna) and then second is Raktika (Madhya), etc.

The Swarsthans or Registers

Swara defined so far in their heptad occur as threefold classified based on their registers defined as Mandra (Lower), Madhyam (Middle) and Tara (High) (39)

Thus far, we have methodically defined Nada, Shruti, and their classifications and also Swara and their classifications and relationships to Shrutis. We now further clarify the Swara (Notes) classification in three registers directly aligned with the three physiological places where the sounds are produced.

1.   Mandra Swara (Lower Register) – Heart

2.   Madhyam Swara (Middle Register) – Throat

3.   Tara Swara (Higher Register) - Head (Brain)

12 modified (vikrit) notes

(vi) द्वादश विकृतस्वराः

त एव विकृतावस्था द्वादश प्रतिपादिताः ॥३९॥

च्युतोऽच्युतो द्विधा षड्जो द्विश्रुतिविकृतो भवेत् ।
साधारणे काकलीत्वे निषादस्य च दृश्यते ॥४०॥

साधारणे श्रुति षाड्जीमृषभः संश्रितो यदा ।
चतुःश्रुतित्वमायाति तदेको विकृतो भवेत् ॥४१॥

साधारणे त्रिश्रुतिः स्यादन्तरत्वे चतुःश्रुतिः ।
गान्धार इति तद्भेदौ द्वौ निःशङ्केन कीर्तितौ ॥४२॥

मध्यमः षड्जवद् द्वेधाऽन्तरसाधारणाश्रयात् ।
पञ्चमो मध्यमग्रामे त्रिश्रुतिः कैशिके पुनः ॥४३॥

मध्यमस्य श्रुति प्राप्य चतुःश्रुतिरिति द्विधा ।
धैवतो मध्यमग्रामे विकृतः स्याच्चतुःश्रुतिः ॥४४॥

कैशिके काकलीत्वे च निषादस्त्रिचतुःश्रुतिः ।
प्राप्नोति विकृतौ भेदौ द्वाविति द्वादश स्मृताः ॥४५॥

ते शुद्धैः सप्तभिः सार्धं भवन्त्येकोनविंशतिः ।

Swaras in their pure form are seven as we have defined earlier. Now we see that in their modified form the Swara number twelve and they are defined as indicated. (39-45)

Notes in their pure form are seven. However, based on their production in practice, if their standard pitch is modified, as it is often the case the notes become "Vikrit" (modified). The Standard pitch is modified by either augmenting it with some Shruti from its closest note on either side of it or by lowering the pitch of the standard tone. From the verse, 40 to 45 above the exact process of such modification is highlighted which we will illustrate now and this results in twelve total notes

with seven pure notes and five modified. The table
below highlights the complete numerical system of the
Pure and Modified Notes i.e. the Shudh (7) and Vikrit
(12) Swara.

Let us now understand how to decipher the evolution of
Modified Notes. We will take each Swara separately
and analyze its Shruti structures in all standard and
modified forms. Rows 1-7 represent the Pure or Shudh
7 Swar. Columns 1 – 10 explain the process of
modification of pure notes into Modified notes and the
corresponding Shruti structure and intervals.

		Shudh and Vikrit Swaras								
		Pure and Modified Notes								
		Std. Notes	Std. Notes	Mod. Form 1	Mod. Form 1	Mod. Form 2	Mod. Form 2	No. of Mod. Swara	Final Swara Shruti	Final Swara Shruti	Final Swara Shruti
		Swara Shrutis	22 Shruti Seq.	Swara Shrutis	22 Shruti Seq.	Swara Shrutis	22 Shruti Seq.		Std. Notes	Mod. Form 1	Mod. Form 2
R1	Shadaj	4	1-4	2	2-3	2	3-4	2	4th	3rd	4th
R2	Rishabh	3	5-7	4	4-7	-	-	1	7th	7th	-
R3	Gandhar	2	8-9	3	8-10	4	8-11	2	9th	10th	11th
R4	Madhyam	4	10-13	2	11-12	2	12-13	2	13th	12th	13th
R5	Pancham	4	14-17	4	13-16	3	14-16	2	17th	16th	16th
R6	Dhaivat	3	18-20	4	17-20	-	-	1	20th	20th	-
R7	Nishad	2	21-22	3	21-1	4	21-2	2	22nd	1st	2nd
		C1	C2	C3	C4	C5	C6	C7	C8	C9	C10

		No. of Mod. Swara		Swara Shrutis	22 Shruti Seq.	Final Swara Shruti	Modification Process Name	Comments
	Swara	Swara						
Shudh and Vikrit Swaras								
Pure and Modified Notes								
R1	Shadaj	2	S	4	1-4	4th	Pure Swar	Pure Swar is composed of 4 Shrutis
			M1	2	2-3	3rd	Chyuta Shadaj Sadharana	1st Shruti of Shudh Shadaj goes to (Lower) Nishad and last 4th Shruti of Shudh goes to Rishabh
			M2	2	3-4	4th	Achyut Shadaj Kakali Nishad formation	First 2 Shrutis of Pure Shadaj go to Mandra Saptak (Lower register) Nishad. Here 4th final Shruti of Shadaj remains same as Pure Shadaj hence it is called Achyut Shadaj and is simply lowered version of Shadaj.
R2	Rishabh	1	S	3	5-7	7th	Pure Swar	Pure Swar is composed of 3 Shrutis
			M1	4	4-7	7th	Chyuta Shadaj Sadharana	Rishabh gains one Shruti from Shadaj becoming 4 Shruti Swar
			-	-	-	-	-	-
R3	Gandhar	2	S	2	8-9	9th	Pure Swar	Pure Swar is composed of 2 Shrutis
			M1	3	8-10	10th	Sadharana Gandhar Chyuta Madhyam Sadharana	Here Gandhar gains first Shruti from Madhyam and hence the name of process.
			M2	4	8-11	11th	AntarGandhar Achyut Madhyam	First 2 Shrutis of Pure Madhyam go to Gandhar. Here 4th final Shruti of Madhyam remains same as Pure Madhyam hence it is called Achyut Madhyam process and is simply lowered version of Madhyam, while Gandhar gains two first of Madhyam creating AntarGandhar.

209

	Shudh and Vikrit Swaras							
	Pure and Modified Notes							
	Swara	No. of Mod. Swara	Mod. Swara	Swara Shrutis	22 Shruti Seq.	Final Swara Shruti	Modification Process Name	Comments
R4	Madhyam	2	S	4	10-13	13th	Pure Swar	Pure Swar is composed of 4 Shrutis
			M1	2	11-12	12th	Chyuta Madhyam Sadharana	Here Gandhar gains first Shruti from Madhyam and Pancham gains the last Shruti of Madhyam, hence the Madhyam is Chyut as its final note Shruti is moved to Pancham and it's hence named in the process.
			M2	2	12-13	13th	Achyut Madhyam AntarGandhar	First 2 Shrutis of Pure Madhyam go to Gandhar. Here 4th final Shruti of Madhyam remains same as Pure Madhyam hence it is called Achyut Madhyam process and is simply lowered version of Madhyam, while Gandhar gains two first of Madhyam creating AntarGandhar
R5	Pancham	2	S	4	14-17	17th	Pure Swar	Pure Swar is composed of 4 Shrutis Pancham where Note Shruti is 17th is the ShadjGram Pancham
			M1	4	13-16	16th	Chyuta Madhyam Sadharana	This is the Modified form of Madhyam Gram Pancham where the Pancham gains final note Shruti of Madhyam
			M2	3	14-16	16th	Madhyam Gram Pancham	This is the Standard form of Three Shruti Madhyam Gram Pancham where final Note Shruti is on 16th Shruti. Here the final note Shruti of Pancham the 17th is given to Dhaivat
R6	Dhaivat	1	S	3	18-20	20th	Pure Swar	Pure Swar is composed of 3 Shrutis
			M1	4	17-20	20th	Madhyam Gram Pancham	Here the final note Shruti of Pancham the 17th is given to Dhaivat when previous Madhyam Gram Pancham is created.
			-	-	-	-	-	-
R7	Nishad	2	S	2	21-22	22nd	Pure Swar	Pure Swar is composed of 2 Shrutis
			M1	3	21-1	1st	Kaishika Nishad Chyuta Shadaj Sadharana	Here Nishad gains one Shruti from the Shadaj which is the final note Shruti of Shadaj.
			M2	4	21-2	2nd	Kakali Nishad Achyut Shadaj	First 2 Shrutis of Pure Shadaj go to Mandra Saptak (Lower register) Nishad. Here 4th final Shruti of Shadaj remains same as Pure Shadaj hence it is called Achyut Shadaj and is simply lowered version of Shadaj.

The above table describes in detail the process of the Twelve Vikrit (Modified) Swars. What is key to understand in studying these relationships is that no

210

matter which modified or pure Swar you analyze, the total number of Shrutis is always 22 and is relative to the addition and subtraction of the modified Shruti numbers as can be seen. E.g. for Chyut Shadaj Sadharana when Shadaj becomes two Shrutis, its extra Two Shrutis have to be assigned to its closest corresponding Swar before and after. One goes to Nishad and one to Rishabh that also is modified simultaneously, thus the natural order of 22 Shrutis is never affected. All of this play of Pure and modified occurs in the context of the 22 Shrutis.

The second observation to note is that when a Pure Swar such as Shadaj is lowered in its number of Shrutis from 4 to 2 as an example, its Pitch remains unaffected but its Tonal content is lowered. Shadaj in pure form is 4 Shrutis away from Nishad, so in its first modified form, it becomes two Shrutis short. However, in the *Indian Shruti and Swar system Tonal value of the Swara is RELATIVE.* It is always relative to the preceding and following Swars and their Shrutis. Therefore, in this case, the Nishad increases in tonal content by one with Rishabh and Shadaj decreases in its Duration of length and hence its tonal content. Yet the Pitch remains the same. So the modification is most definitely felt and experienced yet Pitch remains the same.

Thirdly, we observe in the Table above, in cases of Shadaj, Rishabh, Madhyam and Dhaivat the final Note Shrutis (The Shruti at which the Swara ends and is fully defined) of modified and Pure Swar coincide. This is not a reason to be confused as the Tonal value of the modified Swara is still modified. This is because the Shruti interval and duration of these Swara are different between their Pure and Modified forms in spite of having common Note Shrutis.

Finally, we introduced a new term in the table above which is "Gram" in Shadaj Gram and Madhyam Gram and these will be discussed at length in their own respective chapter in the following section. For this discussion, one needs to note that a Gram is a series of Swars in a certain order of relation. As names suggest Shadaj Gram has Shadaj as its primary note and Madhyam Gram has Madhyam as its primary note. The second most important difference that is of importance to us in this discussion of modified Swars is that Pancham has a very specific definition in both the Gram. In Shadaj Gram Pancham is 4 Shruti interval and its Note Shruti is the 17th Shruti sequentially. In Madhyam Gram Pancham is of Three Shruti Interval and its Note Shruti is the 16th Shruti in the 22 Shruti Sequence. This certainly affects all other notes as well and we will explore that in much detail in the Gram chapter.

Swaras in nature (Birds and animals)

Here we define the relationship of Swars to Nature. All music is based on natural laws of the Universe and amongst living beings, the sounds of the notes correspond to the sounds produced by following flora and fauna who also relate intimately to the seven notes. (46)

(vii) सप्तस्वराणामुच्चारयितारः पशुपक्षिणः
मयूरचातकच्छागक्रौञ्चकोकिलदर्दुराः   ॥४६॥
गजश्च सप्त षड्जादीन्क्रमादुच्चारयन्त्यमी ।

Notes in their pure form are naturally occurring and the ancient sages of Indian music who knew Nature in the deepest intimacy identified the animals corresponding to their sounds with the notes as well as the aesthetic

212

"Rasa" mood of these notes.

	Swar with Corresponding Animals and their "Rasa"					
		1	2	3	4	5
	Swara	Sanskrit Note	English Note	English Name	Nature	"Rasa" Aesthetics
1	Shadaj	Sa	C	Do	Peacock	Heroism Wonder Terror
2	Rishabh	Re	D	Re	Ox - Animal Chataka - Bird	Heroism Wonder Terror
3	Gandhar	Ga	E	Mi	Goat	Compassion
4	Madhyam	Ma	F	Fa	Crane	Humour Love
5	Pancham	Pa	G	Sol	Blackbird	Humour Love
6	Dhaivat	Dha	A	La	Frog	Alert
7	Nishad	Ni	B	Si	Elephant	Strength Compassion

Sonant, Consonant, Dissonant, Assonant notes

(viii) वादि-संवाद-विवादानुवादभदन चतुविधाः स्वराः
चतुर्विधाः स्वरा वादी संवादी च विवादाधपि ॥४७॥
अनुवादी च वादी तु प्रयोगे बहुलः स्वरः ।

श्रुतयो द्वादशाष्टौ वा ययोरन्तरगोचराः ॥४८॥
मिथः संवादिनौ तौ स्तो निगाबन्यविवादिनौ ।
रिधयोरेव वा स्यातां तौ तयोर्वा रिधावपि ॥४९॥
शेषाणामनुवादित्वं वादी राजात्र गीयते ।
संवादी त्वनुसारित्वादस्यामात्योऽभिधीयते ॥५०॥
विवादी विपरीतत्वाद्धीरंहुक्तौ रिपूपमः ।

213

Now we discuss Fourfold Classification of Notes into; Vadi, Samvadi, Vivadi, and Anuvadi (47-50)

1.   **Vadi (Sonant)** – literally means speaker or the note that rules like a King and is most frequent in use. Vadi establishes the melodic foundation of the Raga. This is the fundamental "Tonic" note in the definition of Gram and Jati ragas to be defined later.

2.   **Samvadi (Consonant)** – meaning the corresponding note that converses with the Vadi note above and is second-most in use and is like a Minister. Samvadi creates the dialogue (Samvad) of pleasing delightful melodious structure between itself and Sonant (Vadi)

3.   **Vivadi (Dissonant)** – Note of discord or opposing force like an Enemy of the Vadi note and usually to be avoided. Notes having 2 Shrutis are considered Dissonant or Vivadi per most scholars. Thus, Ga and Ni are Vivadi to all notes in individual usage. In addition, another yardstick is that if the assonant and consonant notes are put together they become dissonant.

4.   **Anuvadi (Assonant)** – Note that sounds afterward and lingers post usage of the Vadi main note. It supports the Sonant note and is akin to a humble servant. This note has no individuality of its own and its sole purpose is to enhance the leading Vadi Swara in

harmony with the consonant note. Notes that are not mutually related either as consonant or dissonant are considered by default to be assonants. E.g., if Rishabh is used for Pancham, Shadaj or Madhyam or if Dhaivat is used for Pancham, Shadaj or Madhyam then these are used as Anuvadi.

Shadaj-Madhyam & Shadaj-Pancham Bhava

The above definition of Fourfold types of Notes now leads us to define the most important relationship of what constitutes most melodious music. Essentially the relationship between Vadi and Samvadi (Sonant and Consonant) Swara is defined as the "Samvad" or the melodious structure and dialogue in the compositions.

Shadaj-Madhyam Bhava - Notes with an interval of 9 Shrutis between them (Shadaj – Madhyam Bhava where Shadaj is Vadi and Madhyam is Samvadi). The relationship of Shruti distance between the final Note Shruti of Shadaj 4th and that of final note Shruti of Madhyam 13th (13-4 = 9) defines this Sonant Consonant pair.

Shadaj-Pancham Bhava - Notes with an interval of 13 Shrutis between them (Shadaj – Pancham Bhava where Shadaj is Vadi and Pancham is Samvadi). The relationship of Shruti distance between the final Note Shruti of Shadaj 4th and that of final note Shruti of Pancham 17th (17-4 = 13) defines this Sonant Consonant pair.

Most Ancient sages such as Bharatmuni, Matang, and Abhinav Gupt all define these Shadaj Madhyam and Shadaj Pancham as 9 and 13 Shruti interval pairs respectively. Sharangdev of Sangeet Ratnakar slightly differs here whereby he DOES NOT count the Note Shruti of starting note into his pair interval hence his pairs are 8 and 12 respectively for Shadaj-Madhyam and Shadaj-Pancham Bhava. We stick to the more commonly accepted 9 and 13 interval definition. Either way, the result of this is that the "Swar Samvad" in Indian music is not just a dialogue of consonance (acoustical phenomena) between Sonant and Consonant, but also a Melodic phenomenon. The more such pairings are used in the musical performance, the more melodious it is in the auditory experience. The table below defines these pairs of Melody.

Note in the below table that the Yellow sections represent the definition of the Vadi Samvadi pairs as permitted by Sharangdev while the non-yellow sections define Vadi Samvadi pairs allowed by other sages' requirement of "SamShrutikata" (Identical measure of Shruti for the two notes to be in pairs. E.g. Sa and Ma as well as Sa and Pa have 4 Shrutis each and hence are allowed). However, as Ma is 4 Shrutis and Ni is 2 Shrutis even though their distance is 13 (25th Shruti - 13). They are not considered in the definition per the identical Shruti requirement. However, Sharangdev would allow them also as he along with another scholar Datilla do not impose that identical Shruti requirement for the pairs and as such any pair meeting 9 or 13 distance would suffice for them.

Seq. Shruti #	Western Swar Notation	Rasa Aesthetics	Sanskrit	Sanskrit Notation	Sa-Ma (9)	Sa-Pa (13)
4	Shadaj C	Heroic Wonder Terror	▯▯▯	▯▯	▯	▯
5	_ D flat				•	•
6	_ B sharp				•	•
7	Komal Rishabh C sharp	Refuge	▯ ▯▯	▯	▯ .	▯
8	Shudh Rishabh D	Fear Refuge	▯ ▯	▯	▯	▯
9	Tivra Rishabh E flat	Disgust	▯▯▯ ▯▯▯	▯.	•	▯ .
10	Komal Gandhar F flat	Very Sad Pathos	▯ ▯▯▯ ▯▯▯	. ▯	•	▯▯
11	_ D sharp			–	•	•
12	Shudh Gandhar E	Sad Pathos	▯ ▯▯▯	▯	▯	▯▯
13	Shudh Madhyam F	Adornment Laughter	▯▯▯▯▯ ▯▯▯▯▯	▯	▯ .	▯▯▯
14	_ G flat			–	•	•
15	_ E sharp			–	•	•
16	Tivra Madhyam F sharp	Adornment Laughter	▯▯▯▯▯ ▯▯▯▯▯	▯ .	▯▯	▯
17	Panchama G	Adornment Laughter	▯▯▯▯▯ ▯▯▯▯▯	▯	▯▯▯	▯
18	_ A flat			–	•	•
19	_ _			–	•	•
20	Komal Dhaivat G sharp	Humility	▯▯▯ ▯▯	▯	▯	•
21	Shudha Dhaivat A	Refuge	▯ ▯▯	▯	▯	▯
22	Tivra Dhaivata B flat	Fear Refuge	▯ ▯	▯ .	▯.	▯
23	Komal Nishad C flat	Pathos	▯ ▯▯▯ ▯	.▯▯	.▯	•
24	_ A sharp			–	•	•
25	Shudha Nishad B	Sad	▯ ▯▯▯	▯▯	▯	▯ .

Other Swara Classifications

(ix) स्वराणां कुल-जाति-वर्ण-देवतानां-च्छन्दोरस कथनम्

गीर्वाणकुलसंभूताः   षड्जगान्धरमध्यमाः ।
पञ्चमः   पितृवंशोत्थो   रिधावृषिकुलोद्भवौ ॥५२॥

निषादोऽसुरवंशोत्थो ब्राह्मणाः समपञ्चमाः   ।
रिधौ तु क्षत्रियौ ज्ञेयौ वंश्यजाती निगौ मतौ ॥५३॥
शूद्रावन्तरकाकल्यौ स्वरौ वर्णास्त्विमे क्रमात् ।
पद्माभः पिञ्जरः स्वर्णवर्णः कुन्दप्रभोऽसितः ॥५४॥
पीतः कर्बुर   इत्येषां जन्मभूमीरथ ब्रुवे ।
जम्बूशाककुशक्रौञ्च   शाल्मलीश्वेतनामसु ॥५५॥
द्वीपेषु पुष्करे चेते जाताः षड्जादयः क्रमात् ।
वह्निनर्बोधाः शशाङ्कश्च लक्ष्मीकान्तश्च नारदः ॥५६॥
ऋषयो ददृशुः पञ्च षड्जादींस्तुम्बुरुर्धनी ।
वह्निब्रह्मसरस्वत्यः   शर्वश्रीशगणेश्वराः ॥५७॥
सहस्रांगुरिति प्रोक्ताः क्रमात्षड्जादिदेवताः ।
क्रमादनुष्टुबगायत्री त्रिष्टुप्च बृहती ततः ॥५८॥
पङ्क्तिरुष्णिक्च जगतीत्याहुश्छन्दांसि सादिषु ।
सरो वीरेऽद्भुते रौद्रे धां बीभत्से भयानके ॥५९॥
कायौ गनी तु करुणे हास्यभृृङ्गारयोर्मपौ ।

Now we discuss various other classifications that define the multitudes of characteristics of the Swara based on the Hindu Vedic traditions. Classification of Notes into; Lineage, Social Status, Color, Continent, Sages, Divinity, Meter, and Aesthetic moods. Summarized in the table below. (52-59)

Swar Vedic Classifications								
		1	2	3	4	5	6	7	8
	Swara	Hindi	English Note	Shrutis	Divinity	Sages	"Rasa" Aesthetics	"Rasa" Bhava	Nature
1	Shadaj	Sa	C	4	Vahni (Fire)	Vahni (Fire)	Heroism Wonder Wrath/Terror	Vira Adbhut Raudra	Peacock
2	Rishabh	Re	D	3	Brahma	Brahma	Heroism Wonder Wrath/Terror	Vira Adbhut Raudra	Ox - Animal Chataka - Bird
3	Gandhar	Ga	E	2	Sarasvati	Sashanka (Moon)	Pathos Compassion	Karun	Goat
4	Madhyam	Ma	F	4	Shiva	Lakshmi kanta (Laxmi Narayana)	Humour Love	Hasya Shringar	Crane
5	Pancham	Pa	G	4	Vishnu	Narada	Humour Love	Hasya Shringar	Blackbird
6	Dhaivat	Dha	A	3	Ganesha	Tumbru	Abhor Fear Alert	Bibhatsa Bhayanak	Frog
7	Nishad	Ni	B	2	Aditya (Sun)	Tumbru	Pathos Compassion	Karun	Elephant

Swar Vedic Classifications								
		1	2	3	4	5	6	7	8
	Swara	Hindi	English Note	Shrutis	Continent	Meter	Lineage	Social Status	Color
1	Shadaj	Sa	C	4	Jambu	Anushtup	Divine	Brahmin	Red
2	Rishabh	Re	D	3	Saka	Gayatri	Sages	Princely	Pale Yellow
3	Gandhar	Ga	E	2	Kusha	Tristup	Divine	Merchant	Golden Yellow
4	Madhyam	Ma	F	4	Krauncha	Brhati	Divine	Brahmin	Sprkling White
5	Pancham	Pa	G	4	Salmali	Pankti	Manes	Brahmin	Black
6	Dhaivat	Dha	A	3	Shvet	Ushnik	Sages	Princely	Plain Yellow
7	Nishad	Ni	B	2	Pushkara	Jagati	Asura	Merchant	Variegated

Process to Purest Swara

So what is the Practice? How can one get closer to the Purest Swars? This occurs in Three Stages.

1. **Attention** – When singing the Attention is not necessarily on Singing, it is on "Listening" because when you concentrate only on words and your speech, you will lose the frame of reference that is needed to attend to for Purity. This attention is the most fundamental quality required to start getting closer in practice to the Pure Swar. Practically speaking that is why a quiet and isolated corner of a location is preferred for practice. Because each Swar must be explored in its fullest form and the singer must try to achieve the closest resemblance to its reproduction from a Harmonium or Sarangi or Vina.

2. **Concentration** – Slowly when you are attentive to the degrees of closeness with the Pure Swar, one will start understanding the moments when we come very close. In those moments, we feel that the Swara we are producing matches the resonance of the instrument accompanying the singing. These moments then automatically force our minds to start concentrating on reproducing such moments. In my earlier book on Science of Rhythm, we explored the nature of Parallel Concentration (Avadhaan). That is the natural leading stage proceeding from Concentration.

3. **Meditation** – The more advanced one gets in their Concentration, we achieve Parallel Concentration and then slowly Meditation. This final Stage of Meditation occurs when the performer is detached in their act of singing from the world that exists outside of them and their focus tends to be ONE with the spirit inside. This then becomes the journey of Devotion and the bridge between Anahat and Aahat Naad is built here. In our Universe, there are many stages of Consciousness. Conciseness is not just what modern science can measure; it exists in many forms, e.g. Botanical Consciousness amongst the natural world of plants and trees and forests, Animal Consciousness amongst the animals of our planet, Human Consciousness amongst all living population of the world. These all are connected via Supramental consciousness connecting all humans to the ultimate Divine Consciousness of the Providence who created the entire Universe. The Meditative state of music helps in traversing these planes of Consciousness.

Section 3 - Gram, Murchana-Krama, Taan

In this Section two of discussing the next group of the Ten Prana of Swara, we now explore in the following order the next set of four pranas Gram, Murchana, Krama, and Taan.

Gram

Gram Definition

ऋथ चतुर्थं ग्राममृर्च्छनाक्रमतानप्रकरणम्

क. ग्रामः

(i) ग्रामलच्चणम्

ग्रामः स्वरसमूहः स्यान्मूर्च्छनाऽऽदे: समाश्रयः ।

Section 4

Grāma, Mūrcchanā, Krama and Tāna

Having discussed the nature and form of individual notes 'Swara' and their relationships to 'Shruti' in the earlier section, we now discuss 'Grama' which essentially is a term denoting a group. In Sanskrit, the word 'Grama' means a village. Here we know the musical Grama as a 'village', a group of related notes.

Individual notes having presented themselves to us, the process of creating melody continues further with the concept of Grama. The individuality of each note essentially shines even further melodically, when presented in a 'set' of related tonality groupings, which we call 'Grama'. Therefore, Grama defines a group of notes having specific ordered relative 'Shruti' values organized to include the integrated set of the heptad 'saptak' seven-note sequence within itself.

Gramas form the foundational basis of tonal 'Shruti' based on organized groupings that define Indian Melody. The question arises as to "Why do we need Grama?" The answer is that Grama further defines more refined systems of formal melodic presentations in music. These terms of groups of further refinement are Murchana, Murchana-series (Krama), note-series (taan), ornamentation (Alankara), graces(gamaka) and archetypes (jatis), etc. These certainly will be discussed at length individually later on but here we mention their names because these subsequent melodic systems are based on foundational groupings of Grama.

In other words, 'Grama' is a group of Swara that have specifically ordered sequences of 'Shrutis' that further constitute melodic conceptual refinements of the terms we just mentioned above such as Murchana, etc. The concept of Grama is the essential foundation that defines what we know as the 'Grama-Murchana' system of the Indian Melodic creative process.

2 main Grams

(ii) द्वौ ग्रामौ (धरातले)

तौ   द्वौ धरातले तन्त्र   स्यात्खड्जग्राम आदिमः ॥१॥
द्वितीयो   मध्यमग्रामस्तयोर्लक्षणमुच्यते ।

In the material world, two Grama are observed to be present namely, Shadaj Grama and Madhyam Gram whose characteristics we will define further. (1)

Notes i.e. 'Swara' in our music can be leveraged in either their pure 'shudh' state or in modified 'vikrit' state. Hence, the two Grama relate to this concept where Shadaj Grama is referencing the 'pure' tones while Madhyam Grama deploys 'modified' form of notes. Further, the condition of a Grama to be in existence is that the 'Swara' in the Grama must be in consonance with each other and have a specific 'Shruti' tonal order. This is possible only in the Grama that starts with Shadaj or Madhyam Swara. Hence, Grams of other notes are not used nor defined. An exception is Gandhar Grama which is employed by the 'Gandharva' the celestial beings and not mortal beings.

Definition of Shadaj Gram and Madhyam Gram

(iii) षड्जमध्यमग्रामयोर्लचणम्

षड्जग्रामः पञ्चमे स्वचतुर्यश्रुतिसंस्थिते ॥२॥
स्वोपान्त्यश्रु तिसंस्थेऽस्मिन्मध्यमग्राम इष्यते ।

Now we understand the definition or rather difference between the two main Grama; Shadaj and Madhyam. The Swara 'Pancham' (P) acts as the main defining principle. In Shadaj Grama 'P' is in its pure form composed of 4 Shrutis. In Madhyam Grama 'P' is modified and is composed of 3 Shrutis with the final 4th 'P' Shruti being taken by the note 'Dhaivat' (D). (2)

3rd Gandhar Gram

(iv) गान्धारग्रामः।

यद्वा धस्त्रिश्रुतिः षड्जे मध्यमे तु चतुःश्रुतिः ॥३॥
रिमयोः श्रुतिमेकैकां गान्धारश्चेत्समाश्रितः ।
षश्रुति धो निषादस्तु घश्रुति सश्रुति श्रितः ॥४॥
गान्धारग्राममाचष्ट तदा तं नारदो मुनिः ।
प्रवर्तंते स्वर्गलोके ग्रामोऽसौ न महीतले ॥५॥

Shri Naradamuni (one of the founding Acharyas of the Swara Shastra) defines Gandhar Gram as one that is used by divine beings and not to be used by mortals. Gandhar Grama comes into existence by following modifications relative to the 'Pure' Shadaj Grama – Gandhar takes one shruti from Rishabh and Madhyama each. Dhaivat takes one Shruti of Pancham and Nishad takes one shruti each from Dhaivat and Shadaj (3, 4, 5)

The table summary below assists us in understanding the specific order of Shrutis and their Swara in the definition of each of the three Grama.

Grama Shruti Swara										
Grama Name	**Swara & Shruti No.**									
	S	R	G	M	P	D	N	S	R	G
Shadaj Grama	4	3	2	4	4	3	2			
Madhyam Grama				4	3	4	2	4	3	2
Gandhar Grama			4	3	3	3	4	3	2	

Note each Grama starts on its corresponding Swara

Justification of the Three Gram

(v) ग्रामत्रयस्य नामस्वराणां वैशिष्ट्यम्

षड्जः   प्रधानमाद्यत्वादमात्याधिक्यतस्तथा ।
ग्रामे   स्यादविलोपित्वान्मध्यमस्तु   पुरःसरः ॥६॥
एतत्कुलप्रसूतत्वाद् गन्धारोऽप्यग्रणीर्दिवि ।

The emotional sensibilities behind the names of three
Grama are discussed now. Shadaj being the 'First'
prime note incorporates within it all other notes and
hence the pure Grama of Shadaj is named after it.
Madhyam cannot be omitted as it is in the middle of the
scale and must be included in all melodic compositions
and hence it constitutes the second Grama. Finally,
Gandhar belongs to the same family of divinity who
preside over Shadaj and Madhyam and hence this forms
the third but divine Grama. (6)

Gram and their presiding Deities

(vi) ग्रामाणां देवा।

क्रमाद्   ग्रामत्रये   देवा   ब्रह्माविष्णुमहेश्वराः ॥७॥

The Divinity associated with each Grama is Brahma
(Shadaj), Vishnu (Madhyam) and Mahesh (Gandhar) (7)

Gram and their periodic times of preference

(vii) ग्रामाणां गानकालनियमः

हेमन्तग्रीष्मवर्षासु   गातव्यास्ते   यथाक्रमम् ।
पूर्वाह्णकाले मध्याह्ने ऽपराह्ने ऽभ्युदयार्थिभिः ॥८॥

The observed sensibilities of the groups of notes and their corresponding times and seasons are documented herein. These are not rigid but rather define some suggested themes for the further melodic structures that can be derived from the Grama. The table below summarizes these. (8)

Grama Sensibilities			
Grama Name	Presiding Deity	Optimal Season	Optimal Time
Shadaj Grama	Brahma	Winter	Morning
Madhyam Grama	Vishnu	Summer	Noon
Gandhar Grama	Mahesh	Rain	Evening

Note: As is the case with many aspects of logical and scientific systems, the perceptive emotional sensibilities of those systems are only understood through experience. The Ancient sages in describing the Grama and their sensibilities above are NOT suggesting literal rigidity of time and seasons but rather these are to be used as creative catalysts for future Melodies that evolve from each Grama.

To summarize, the purpose of the Grama is to enable and provide a formal foundation for the organization of Swara, Shruti, Moorchna, Tana, Jati, and Raga. This forms the foundational system of Indian melodic structure creation called Grama-Murchana-Jati system. Therefore, if we look at the last element of these concepts, the Raga, which derives from a Jati, these times, and seasonality aspects of Grama essentially provide some reference feelings of sense in the tonal structures of melody.

Murchana - Krama

Murchana Definition

ख. मूर्च्छना: क्रमाश्च

(i) मूर्च्छनालक्षणम्

क्रमात्स्वराणां सप्तानामारोहश्चावरोहणम् ।
मूर्च्छनेत्युच्यते ग्रामद्वये ताः सप्त सप्त च ॥९॥

The cyclical ascending and descending movement of the seven notes in successive order in the heptad (saptak) is called Moorchna. There are seven Murchanas in each of the two Gramas (Shadaj and Madhyam) employed in the material world. (9)

Having established the foundation of further melodic refinements earlier in the definition of Grama, we now explore Murchana, which is the next element of the process of creating Melody in the Indian system of music. The word Murchana is derived from Sanskrit root word 'moorch' which means simultaneously two opposing ideas; one is 'to faint' (fall down) and it means 'to increase' (to rise). This would be clear as we study natural law below.

"Ascendant and Descendent" in nature – When we study the physically manifested world around us we cannot escape the fact of nature that "everything" in the universe is self-controlled through a cyclical movement of rising and falling. This law of Ascent and Descent is omnipresent all the way starting from Sunrise to Sunset, Waxing and Waning Moon that causes the Rising and falling tides. Over long time span, the physical form of

229

the earth itself rises and falls, mountains become seabed and from oceans rise new Mountains. North becomes south and vice versa. These movements are essentially the 'breath' of the Supreme Brahman itself; exhale and inhale, creation and destruction and infinite cycle of new 'conscience' evolving in the physically manifested world.

It is this precise yin and yang of Ascent and Descent movement of the 7 notes in successive order that define Murchana and give rise to what we know in practical terms as practiced melodic structures called 'Raga'. In other words, the act by which the practical melodic structures of Raga are formed is known as Murchana. Without the movement of Murchana, there can be no Raga. Without the act of 'going' up and 'coming' back down, there can be no musical journey and hence this is the root sensibility to understand about Murchana.

A soul is always going and arriving somewhere. *Thus, Music itself is akin to a Universal Pilgrimage.* The question is 'who' is on this universal pilgrimage, in context of creativity of our melody; the pilgrims are the seven "Swara" themselves in the constant act of moving up and down. The function of the Murchana is thus to develop and expand upon the melodic structure of Raga.

14 Murchana and Tonality

(ii) षड्जमध्यमग्रामस्थमूर्च्छनानां संज्ञाः

षड्जे तुत्तरमन्द्रा ऽऽदौ रजनी चोत्तरायता ।
शुद्धषड्जा मत्सरीकृदश्वक्रान्ता ऽभिरुद्गता ॥१०॥
मध्यमे स्यात्तु सौबीरी हारिणाश्वा ततः परम् ।
स्यात्कलोपनता शुद्धमध्या मार्गी च पौरवी ॥११॥
हृष्यकेति,

(iii) मूर्च्छनानामारम्भकस्वराः

, अथ तासां तु लक्षणं प्रतिपाद्यते ।
मध्यस्थानस्थषड्जेन मूर्च्छना ऽऽरभ्यते ऽग्रिमा ॥१२॥

अधस्तन्नॆनिषादाद्यैः षडन्या मूर्च्छनाः क्रमात् ।
मध्यमध्यममारभ्य सौबीरी मूर्च्छना भवेत् ॥१३॥
षडन्यास्तदधोऽधःस्थस्वरानारभ्य तु क्रमात् ।
षड्जस्थानस्थितैन्र्याद्यैं रजन्याद्याः परे विदुः ॥१४॥
हारिणाश्वादिका गाढ्यॆमंध्यमस्थानसंस्थितः ।
षड्जादीन्मध्यमादींश्च तदूर्ध्वं सारयेत्क्रमात् ॥१५॥

The Shadaj Grama and Madhyam Grama each have seven Murchana and these in total form the fourteen Murchana named in the table below. The first Murchana in each grama begins with the Shadaj and Madhyam respectively of the 'middle' register. All the six Murchanas in each Grama then are defined by successively placing one swara below the other in their regular order. (10-15)

Moorchana and two Gramas

Shadaj Grama Moorchana — Swara Structures

		1	2	3	4	5	6	7
1	Uttara Mandra	s	r	g	m	p	d	n
2	Rajani	n	s	r	g	m	p	d
3	Uttarayata	d	n	s	r	g	m	p
4	Suddha Shadaj	p	d	n	s	r	g	m
5	Matsarikrita	m	p	d	n	s	r	g
6	Ashvakranta	g	m	p	d	n	s	r
7	Abhirudgata	r	g	m	p	d	n	s

Madhyam Grama Morrchana

		1	2	3	4	5	6	7
1	Sauviri	m	p	d	n	s	r	g
2	Harinashva	g	m	p	d	n	s	r
3	Kalopanata	r	g	m	p	d	n	s
4	Shuddha Madhyam	s	r	g	m	p	d	n
5	Margi	n	s	r	g	m	p	d
6	Pauravi	d	n	s	r	g	m	p
7	Hrishyaketi	p	d	n	s	r	g	m

We should note that since the earliest documented treatise on Indian music by Bharat in his Natya Shastra, it is specified that for Murchanas to be able to span all the three registers of notes the middle point is defined by the middle register. Therefore, if Murchana means moving up and down the registers of swara then this movement is with respect to the middle and hence this middle register forms the basis of first Murchana and everything evolves from that. Therefore, the Murchana of Shadaj Grama and Madhyama Grama begin with the Shadaj and Madhyam swara of the middle register.

It is important to note that while Swara structures in both Grama sometimes might be the same in some Murchana, their tonality is completely different and hence they are still different in their melody. This is because the Pancham in Madhyam gram is composed of 3 shrutis whereas it is made up of four shrutis in Shadaj grama. Thus due to shruti tonality differences in each Grama each of the seven Murchana of both grama have their own unique identity.

The ideal instrument to practice and observe the creation of Murchanas is vina. Depending on the type of vina, Murchana can be obtained across the range of three registers in a vina that has 21 strings tuned to three registers individually. This type of vina is called 'mattakokila' vina. The table below highlights Murchana definition using such a vina. This is clear visually in a table depicting Murchanas and the three registers.

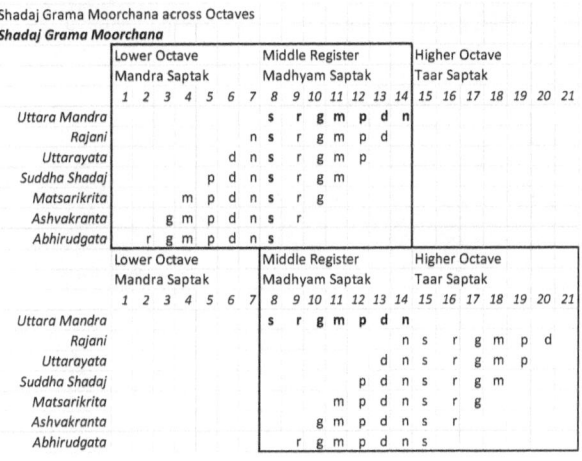

Shadaj Grama Moorchana across Octaves

Shadaj Grama Moorchana

	Lower Octave — Mandra Saptak							Middle Register — Madhyam Saptak							Higher Octave — Taar Saptak						
	1	2	3	4	5	6	7	8	9	10	11	12	13	14	15	16	17	18	19	20	21
Uttara Mandra								s	r	g	m	p	d	n							
Rajani							n	s	r	g	m	p	d								
Uttarayata						d	n	s	r	g	m	p									
Suddha Shadaj					p	d	n	s	r	g	m										
Matsarikrita				m	p	d	n	s	r	g											
Ashvakranta			g	m	p	d	n	s	r												
Abhirudgata		r	g	m	p	d	n	s													

	Lower Octave — Mandra Saptak							Middle Register — Madhyam Saptak							Higher Octave — Taar Saptak						
	1	2	3	4	5	6	7	8	9	10	11	12	13	14	15	16	17	18	19	20	21
Uttara Mandra								s	r	g	m	p	d	n							
Rajani														n	s	r	g	m	p	d	
Uttarayata													d	n	s	r	g	m	p		
Suddha Shadaj												p	d	n	s	r	g	m			
Matsarikrita											m	p	d	n	s	r	g				
Ashvakranta										g	m	p	d	n	s	r					
Abhirudgata									r	g	m	p	d	n	s						

Another technique is to use the middle register strings and tune the 7 strings of that register according to the Murchanas per their shruti intervals. Hence, shruti intervals are needed precisely to be defined in this second approach of creating Murchanas on vina where we use only the middle register. 7 strings are tuned precisely in shruti combinations of a particular Murchana. The same table above is enhanced depicting shruti tonalities as well to understand this idea.

Shadaj Grama Shruti Tonality in Middle Register

Shadaj Grama Moorchana		Middle Octave Register						
		8	9	10	11	12	13	14
1	*Uttara Mandra*	4 s	3 r	2 g	4 m	4 p	3 d	2 n
2	*Rajani*	2 n	4 s	3 r	2 g	4 m	4 p	3 d
3	*Uttarayata*	3 d	2 n	4 s	3 r	2 g	4 m	4 p
4	*Suddha Shadaj*	4 p	3 d	2 n	4 s	3 r	2 g	4 m
5	*Matsarikrita*	4 m	4 p	3 d	2 n	4 s	3 r	2 g
6	*Ashvakranta*	2 g	4 m	4 p	3 d	2 n	4 s	3 r
7	*Abhirudgata*	3 r	2 g	4 m	4 p	3 d	2 n	4 s

Numbers preceding the swara signify the shruti interval.

4 Fold Classification of Murchana

Murchanas we defined earlier are further expanded and classified into four groupings, viz. Shuddha (pure), Sakakali (kakali inclusive), Santara (antara inclusive) and Ubhaygata (inclusive of both kakali and antara). Thus, we form totally 56 Murchanas (7x 4 for Shadaj Grama and 7x 4 for Madhyam Grama) (16-17)

(iv) मूर्च्छना-भेदाः

चतुर्धा ताः पृथक् शुद्धाः काकलीकलितास्तथा ।
सान्तरास्तद्द्वयोपेताः षट्पञ्चाशदितीरिताः ॥१६॥

श्रुतिद्वयं चेत्खड्जस्य निषादः संश्रयेत्तदा ।
स काकली, मध्यमस्य गान्धारस्तद्वन्तरः स्वरः ॥१७॥

Inclusion of Kakali and Antara or combination thereof is the main determining factor in murchana classification. Both Kakali and Antara are referenced in the Swara section earlier and we refresh our understanding of them again here at this key point. Swara are either pure (Shudh) or modified (Vikrit). When the shruti intervals of the pure swaras are modified, they result in the Vikrit swara. Both Kakali and Antara are vikrit in nature.

Kakali Nishad Definition - Pure Nishad is constituted of 2 shrutis and occurs at 22nd Shruti. However if this pure Nishad takes on two extra Shrutis from its next subsequent Shadaj of next register, then the Nishad is extended to 24th Shruti and is made up of 4 shrutis in total. This is the modified Nishad and is known simply as Kakali.

Antara Gandhar Definition – Pure Gandhar also has 2 shrutis in its construction. When it acquires two extra shrutis from the next subsequent Madhyam, this modified Gandhar elongates to become four shruti in its tonality. This is the Antara Gandhar simply known as Antara.

This now positions us to precisely define the four-murchana classifications.

1. **Shudh (pure or standard) Murchana** – These are composed of all pure notes or shudh swara. The standard pure shruti relationships of seven notes are used here viz. 4-3-2-4-4-3-2 (in terms of shruti intervals) or 4-7-9-13-17-20-22 (in terms of sequential 22 shruti numbering)

2. **Sakakali (Kakali Murchana)** – This type includes the Kakali swara in its construct. Instead of pure, Nishad Kakali Nishad is used here. The shruti relation here is as follows; 4-7-9-13-17-20-2 (in terms of sequential 22+2 shruti numbering, the Nishad takes 2 extra shrutis from Shadaj of next register and hence ends on 2)

3. **Santara (Antara Murchana)** – This type includes the Antara swara in its construct. Instead of pure Gandhar, Antara Gandhar is used here. The shruti relation here is as follows; 4-7-11-13-17-20-22 (in terms of sequential 22 shruti numbering, the Antara takes 2 extra shrutis from Madhyam next to it and hence ends 11 instead of 9)

4. **Ubhaygata (Kakali Antara Murchana)** – Here both Kakali AND Antara are used in the Murchana and hence the shruti relationship is of 24 shrutis as follows; 4-7-11-13-17-20-2 (in terms of sequential 22+2 shruti numbering, the Nishad takes 2 extra shrutis from Shadaj of next register and hence ends on 2 and the Antara takes 2 extra shrutis from Madhyam next to it and hence ends 11 instead of 9)

For Shadaj Grama the above example grouping definitions apply for the FIRST murchana. Similarly, one can analyze and apply the four groups to all seven Murchana of Shadaj Grama resulting in 7x4 = 28 Murchana of Shadaj Grama. The only difference in Madhyam Gram is that the Shruti intervals of Pancham change as Pancham is of 3 shrutis in Madhyam Grama so it will occur on 16th Shruti in Madhyam Grama. One can also arrive at 28 Murchana of Madhyam Gram like this and thus obtain the 54 unique murchanas.

Murchana Kramas

The positions of Shadaj and Madhyam in the murchana determine the numbering of the seven murchanas in each grama. E.g. In first murchana of Shadaj grama Shadaj is in the first place and in 4th murchana of Shadaj grama it is in 4th place and so on (18)

(v) मूर्च्छनासङ्ख्या-परिज्ञानोपायः

यस्यां यावतिथी षड्जमध्यमौ ग्रामयोः क्रमात् ।
मूर्च्छना तावतिथ्येव सा निःशङ्केन कीर्तिता ॥१८॥

237

(vi) मूर्च्छना-क्रमाः, तेषां सङ्ख्या च

प्रथमादिस्वरारम्भादेकंका सप्तधा भवेत् ।
तासूच्चार्यान्त्यस्वरांस्तान्पूर्वानुच्चारयेत्क्रमात् ॥१३॥
ते क्रमास्तेषु संख्या स्याद् द्वानवत्या शतत्रयम् ।

Having understood the numbering of seven basic murchanas, we now see the sequential order of all the 54 murchanas. This is called the 'Krama'. Each of the 54 murchanas becomes 7-fold when one starts from the first note of the murchana and ends on the seventh note. Then sequentially starts from the next one of the sequential seven swars as its initiating first note. In this manner, we have a creative span of 54x7 = 392 total Krama of Murchana. (19)

This is the paramount creative idea in the Science of Melody (Swara Shastra). The 392 creative sequences or kramas of murchana are arrived at WITHOUT changing the scale. The basic scale is unchanged and the Murchana does not change nor does it total Shruti values, but what changes is the SEQUENCE of the order of swara in the murchana. We have a total of seven swaras so if we change the order of those seven swars, each of the 54 murchanas now allow us to create seven variations resulting in the 7 kramas (sequences) of the Murchana.

The only difference is that when we consider the murchana from the main 54 groups, they are able to both go up and down in tonality. Whereas the *kramas which are the sub-varieties of the 54* derived by *changing the order of the initial swara (and cyclically going to subsequent 7 swaras)* **are only formed in Ascending order of notes.** This is a very important point to remember when creatively using murchana.

So to summarize, EACH of the 54 murchana has 7 variations of its own (krama) which are formed by starting the murchana with 1ˢᵗ, 2ⁿᵈ, 3ʳᵈ, to the 7ᵗʰ swara notes keeping Shruti value of notes unchanged. What changes is the sequential cyclical order of the initial and ending notes of the murchana. These are the 7 series or Murchana Kramas of the main 54 murchanas resulting in total sub-varieties of 54x7 = 392.

Murchana Deities and Naming

The spiritual aesthetics of the main 7 murchanas are associated with their corresponding divinity as follows. Sage Narada also defines names of each murchana slightly differently in his most ancient of the musical treatise (Naradiya Shiksha) and those names are listed here for reference as well. (20-21)

(vii) मूर्च्छनानां देवताः

यक्षरक्षोनारदाब्जभवनागाश्विपाशिनः   ॥२०॥
षड्जग्रामे मूर्च्छनानामेताः स्युर्देवताः   क्रमात् ।

ब्रह्मेन्द्रबायुगन्धर्वसिद्धद्रुहिणभानवः   ॥२१॥
स्युरिमा   मध्यमग्राममूर्च्छनादेवताः   क्रमात् ।

(viii) नारदकथितानि मूर्च्छना-नामानि

तासामन्यानि नामानि नारदो मुनिरब्रवीत् ॥२२॥

मूर्च्छनोत्तरवर्णा ऽऽद्या षड्जग्रामे ऽभिरुद्गता ।
अभ्यक्रान्ता च सौवीरी हृष्यका चोत्तरायता ॥२३॥

रजनीति समाख्याता ऋषीणां सप्त मूर्च्छनाः ।
आप्यायनी विश्वकृता चन्द्रा हेमा कर्पर्दिनी ॥२४॥

मैत्री चान्द्रमसी पित्र्या मध्यमे मूर्च्छना इमाः ।
नन्दा विशाला सुमुखी चित्रा चित्रवत्तो सुखा ॥२५॥

आलापा चेति गान्धारग्रामे स्युः सप्त मूर्च्छनाः ।
ताश्च स्वर्गे प्रयोक्तव्या विशेषात्तेन नोदिताः ॥२६॥

Sage Narada's definition of names is given in the table below. (22-26)

	Moorchana and two Gramas								Presiding Deities	Sage Narada's Nomenclature
	Shadaj Grama Moorchana	Swara Structures								
		1	2	3	4	5	6	7		
1	Uttara Mandra	s	r	g	m	p	d	n	Yaksha	Uttara varna
2	Rajani	n	s	r	g	m	p	d	Raksha	Abhirudgata
3	Uttarayata	d	n	s	r	g	m	p	Narada	Ashvakranta
4	Suddha Shadaj	p	d	n	s	r	g	m	Brahma	Sauviri
5	Matsarikrita	m	p	d	n	s	r	g	Naga	Hrshyaka
6	Ashvakranta	g	m	p	d	n	s	r	Ashvin	Uttarayata
7	Abhirudgata	r	g	m	p	d	n	s	Varuna	Rajani
	Madhyam Grama Morrchana									
1	Sauviri	m	p	d	n	s	r	g	Brahma	Apyayani
2	Harinashva	g	m	p	d	n	s	r	Indra	Vishvakrita
3	Kalopanata	r	g	m	p	d	n	s	Vayu	Chandra
4	Shuddha Madhyam	s	r	g	m	p	d	n	Gandharva	Hema
5	Margi	n	s	r	g	m	p	d	Siddha	Kapardini
6	Pauravi	d	n	s	r	g	m	p	Shiva	Maitri
7	Hrishyaketi	p	d	n	s	r	g	m	Sun	Chandramasi

Taan

Thus far, we have covered the five characteristics of Swara Shastra; Nada, Shrutis, Swara, Grama and Murchana in detail. Their meanings and their practical measures in physically quantifying the terms, as well as melodically arriving at some form of musical constructs, have been discussed. Now we delve into how these ideas are further expanded upon using the concept of Taan to arrive at formal melodic expansion and subsequently leading to the creation of what we know as structured systems of melody in Indian music called Jati - Ragas.

The word Tana or Taan is derived from Sanskrit root 'Tan', which means to spread, expand, stretch or enlarge and develop the physical form. In the musical context of melody, these Taan constructs are a musical phrase in specific order sung with the notes of swars or words of poetry and expand the melodic content of the performance.

ग. तानाः

(i) शुद्धताननिर्माणविधिः, सङ्ख्या च

1. शुद्धतानलचणम्

तानाः स्पुर्मूर्च्छनाः शुद्धाः षाडवौडुवितीकृताः ।

Shudh Taans

Shudh Taan is formed by taking a Shudh Murchana and drooping either one note from it, which forms Shadav (Hexatonic) Taan and two notes, which forms Audav (Pentatonic) Taan. (27)

Definition of Shudh Taan

Taan and Murchana are both qualified by the word Shudh to denote pure or standard characteristics of that Murchana or Taan. The exact definition of Shudh Taan is that it is a specific note series formed by dropping either one or two notes from the Shudh Murchana. In practical usage of Taan the word Shudh is not used often and just Taan is used to signify a Hexatonic (Shadav) or Pentatonic (Audav) note series derived from dropping one or two notes of a Standard Murchana. The word 'specific' note series is used to describe Shudh Taan because it is without any change in sequential order of notes of the original 7 Shudh Murchana. Whereas in the original 7 Murchana the order of the notes can change, in the Taan the order is NOT changed only the notes are dropped from the original 7 Shudh Murchana.

2. षाडवतानाः

षड्जगाः सप्त   हीनाश्चेत्क्रमात्सरिपसप्तमेः ॥२७॥
तदा ड्ष्टाविंशतिस्ताना मध्यमे सरिगोज्झिताः ।
सप्त क्रमाछ्दा तानाः स्युस्तदा त्वेकर्विंशतिः ॥२८॥
एते   चैकोनपश्चाशद्भये   षाडवा   मताः ।

49 Shadav Taan (Hexatonic series)

There are 49 (28 + 21) Shadav Taans.

Shadaj Grama – the seven murchana of this grama give rise to 28 Shadav Taan (Hexatonic note series). By dropping one note each s, r, p and n in their respective turns in the seven murchanas we get 7x4 = **28 Shadaj Grama Shadav Taan.**

Madhyam Grama - the seven murchana of this grama gives rise to 21 Shadav Taan. By dropping one note each s, r and g in their respective turns in the seven murchanas we get 7x3 = **21 Madhyam Grama Shadav Taan**. (27-29)

This is a great place to refresh that Shudh Murchana are those that do NOT contain any modified notes such as Kakali or Antara but the order of their notes can change. Shudh Taans (simply known as Taan) also do not contain any modified Swara such as Kakali or Antara but also their order of swara in the note-series is 'Specific' and does not change from regular order.

Below table shows this with an example of omitting Pancham from Shadaj Gram and Gandhar from Madhyam Grama to create 7 Shadav Taans in each Grama.

Creation of Shadav Taan Example

		Shudh Murchana							Omission of Pancham Shadav (Hexatonic) Taan					
	Shadaj Grama Moorchana	1	2	3	4	5	6	7	1	2	3	4	5	6	7
1	Uttara Mandra	s	r	g	m	p	d	n	s	r	g	m	_	d	n
2	Rajani	n	s	r	g	m	p	d	n	s	r	g	m	_	d
3	Uttarayata	d	n	s	r	g	m	p	d	n	s	r	g	m	_
4	Suddha Shadaj	p	d	n	s	r	g	m	_	d	n	s	r	g	m
5	Matsarikrita	m	p	d	n	s	r	g	m	_	d	n	s	r	g
6	Ashvakranta	g	m	p	d	n	s	r	g	m	_	d	n	s	r
7	Abhirudgata	r	g	m	p	d	n	s	r	g	m	_	d	n	s
	Madhyam Grama Morrchana								Ommision of Gandhar						
1	Sauviri	m	p	d	n	s	r	g	m	p	d	n	s	r	_
2	Harinashva	g	m	p	d	n	s	r	_	m	p	d	n	s	r
3	Kalopanata	r	g	m	p	d	n	s	r	_	m	p	d	n	s
4	Shuddha Madhyam	s	r	g	m	p	d	n	s	r	_	m	p	d	n
5	Margi	n	s	r	g	m	p	d	n	s	r	_	m	p	d
6	Pauravi	d	n	s	r	g	m	p	d	n	s	r	_	m	p
7	Hrishyaketi	p	d	n	s	r	g	m	p	d	n	s	r	_	m

Note: s, r, p and n can be ommited in Shadaj Grama

Note: s, r and g can be ommited in Madhyam Grama

35 Audav Taan (Pentatonic series)

3. औडुवतानाः

सपाभ्यां द्विश्रूतिभ्यां च रिपाभ्यां सप्त वर्जिताः ॥२९॥
षड्जग्रामे   पृथक्तानाः   एकविंशतिरौडुवाः ।
रिधाभ्यां द्विश्रूतिभ्यां च मध्यमग्रामगास्तु ते ॥३०॥
हीनाश्चतुर्दशैव स्युः पञ्चत्रिंशत्तु ते युताः ।

There are a total 35 (21 + 14) Audav Taans.

Shadaj Grama – the seven murchana of this grama give rise to 21 Audav Taan (Pentatonic note series). By dropping TWO notes each s-p, g-n and r-p in their respective turns in the seven murchanas we get 7x3 = **21 Shadaj Grama Audav Taan**.

Madhyam Grama - **the seven murchanas of this grama give rise to 14 Audav Taan. By dropping TWO note each r-d and g-n in their respective turns in the seven murchanas we get 7x2 =** 14 Madhyam Grama Audav Taan. **(29-30)**

4. षाडबौडुवतान-मिश्रितसङ्ख्या

सर्वं चतुरशीतिः स्युर्मिलिताः षाडबौडुवाः ॥३१॥

Thus, the total number of Shudh Taan is 84 (49+35).

49 (28 Shadaj Grama + 21 Madhyama Grama) Shadav Taans

35 (21 Shadaj Grama + 14 Madhyama Grama) Audav Taans

Below table shows this with an example of omitting the s-p pair from Shadaj Gram and g-n pair from Madhyam Grama to create 7 Audav Taans in each Grama.

Creation of Audav Taan Example

		Shudh Murchana							Omission of s-p Audav (Pentatonic) Taan						
	Shadaj Grama Moorchana	1	2	3	4	5	6	7	1	2	3	4	5	6	7
1	Uttara Mandra	s	r	g	m	p	d	n	x	r	g	m	_	d	n
2	Rajani	n	s	r	g	m	p	d	n	x	r	g	m	_	d
3	Uttarayata	d	n	s	r	g	m	p	d	n	x	r	g	m	_
4	Suddha Shadaj	p	d	n	s	r	g	m	_	d	n	x	r	g	m
5	Matsarikrita	m	p	d	n	s	r	g	m	_	d	n	x	r	g
6	Ashvakranta	g	m	p	d	n	s	r	g	m	_	d	n	x	r
7	Abhirudgata	r	g	m	p	d	n	s	r	g	m	_	d	n	x
	Madhyam Grama Morrchana								Ommision of g-n						
1	Sauviri	m	p	d	n	s	r	g	m	p	d	x	s	r	_
2	Harinashva	g	m	p	d	n	s	r	_	m	p	d	x	s	r
3	Kalopanata	r	g	m	p	d	n	s	r	_	m	p	d	x	s
4	Shuddha Madhyam	s	r	g	m	p	d	n	s	r	_	m	p	d	x
5	Margi	n	s	r	g	m	p	d	x	s	r	_	m	p	d
6	Pauravi	d	n	s	r	g	m	p	d	x	s	r	_	m	p
7	Hrishyaketi	p	d	n	s	r	g	m	p	d	x	s	r	_	m

Note: s-p, r-p, g-n can be ommited in Shadaj Grama
Note: r-d and g-n can be ommited in Madhyam Grama

Kuta Taans

1. कूटतानलक्षणम्

असम्पूर्णाश्च सम्पूर्णा व्युत्क्रमोच्चारितस्वराः ।
मूर्छनाः कूटतानाः स्युः,

Complete or incomplete murchanas with their notes produced in permutation format (non-sequential) create permutational note series known as 'Kuta Taan'.

Definition of Kuta Taan

Let us refresh here that Murchanas are classified in four groupings based on their swara; Shudh, Kakali, Antara and Kakali-Antara.

These definitions assume a 'complete' format of murchanas that include ALL 7 swara in them. In addition, we note that in Shudh murchana the swara can ascend and descend as well but in the murchana-series that are derived from the main 7 (called Kramas), the notes are in ascending order only.

Now if the Murchanas are 'incomplete' that is they have less than seven notes then they still are able to create Taans. We have seen that when one or more notes are omitted from a Murchana it becomes a Taan. Thus, 'incomplete' Murchana by the very definition of omission of notes gives rise to Taan of some form.

The key point to note is that a murchana whether complete or incomplete, creates a 'Kuta Taan' permutational note-series, *when the natural order of its notes is changed*. In this context, a Kuta Taan can also have a descending movement of notes whereas the original murchana-series (Krama) derivations are only ascending. If the natural order of notes is maintained in the Taan then it is by definition Shudh Taan (simply known as Taan) which we discussed previously.

Purna Kuta Taan (Complete)

2. पूर्णकूटतानानां सङ्ख्या

तत्सङ्ख्यामभिदध्महे ॥३२॥

पूर्णाः पञ्च सहस्राणि चत्वारिंशद्युतानि तु ।
एकैकस्यां मूर्च्छनायां कूटतानाः सह क्रमैः ॥३३॥
षट्पञ्चाशन्मूर्च्छनास्थाः पूर्णाः कूटास्तु योजिताः ।
लक्षद्वयं सहस्राणि द्व्यशीतिर्द्वे शते तथा ॥३४॥
चत्वारिंशच्च सङ्ख्याता

Inclusive of the seven, murchana-series (kramas) there are 5040 total permutational note-series 'Kuta Tana' in EACH of the 56 murchanas. Giving rise to the total number of 'Kuta Tana' as **56x5040 = 282,240**

(32-34)

5040 is arrived using permutational mathematics as follows;

Number of Notes	Possible Permutational Series
1	1x1 = 1
2	1x2 = 2
3	2x3 = 6
4	6x 4 = 24
5	24 x 5 = 120
6	120 x 6 = 720
7	720 x 7 = *5040*

Examples with 2 and 3 note 'Kuta Tana' series

2 Notes Kuta Tana with notes s and r = s r, r s
3 Notes Kuta Tana with notes s r g = s r g, r s g (g fixed)
3 Notes Kuta Tana with notes s r g = s g r, g s r (r fixed)
3 Notes Kuta Tana with notes s r g = r g s, g r s (s fixed)

One should note that in forming the 'Kuta Tana' attempt is made to keep the last note fixed for as long as the permutations allow and then switch to the next note as seen in example above.

Mathematical principle of permutation dictates the number of times a given note can remain 'fixed' in the formation of the permutational series. *The number of permutations of preceding series = the number of times each note can remain 'fixed' in the succeeding series.* In our example, the permutations of the first series with 2 notes are 1 x 2 = 2. So in the next succeeding permutation with three notes each of the three notes can be kept 'fixed' twice leading us to 2 x 3 = 6 permutations of Kuta Taans.

Apurna Kuta Taan (Incomplete)

३. अपूर्णकूटतान-निर्माणम्

,अथापूर्णान्प्रचक्ष्महे ।

एकैकान्त्यान्त्यविरहाद्द्दाः षट् षट्स्वरादयः ॥३५॥

एकस्वरो ऽत्र निर्भेदो ऽप्युक्तो नष्टादिसिद्धये ।

क्रमा अकूटतानत्वे ऽप्युक्तास्तेषूपयोगिनः ॥३६॥

Incomplete 'Kuta Taan' as the name suggests does NOT have full seven notes. By dropping the last note from the complete variety sequentially, we get six types of incomplete permutational note-series. Viz. Hexatonic, Pentatonic, Tetratonic, Tritonic, Bitonic and Monotonic series. (35-36)

Thus by dropping the seventh note we get Hexatonic Kuta tana, by dropping sixth and seventh notes we get Pentatonic series and so on and so forth till we drop all six notes and remain with a single note of a monotonic tana series.

4. अपूर्ण कूटतानानां सङ्ख्या

स्युः षाडवानां विंशत्या सह सप्त शतानि तु ।
औडुवानां तु विंशत्या सहितं शतमिष्यते ॥३७॥
चतुःस्वराणां कूटानां चतुर्विंशतिरीरिताः ।
त्रिस्वराः षड् द्विस्वरौ द्वावेकस्त्वेकस्वरो मतः ॥३८॥

Shadav Kuta Tana (Hexatonic) = 720 (which includes original murchana series kramas as well.)

Audav Kuta Tana (Pentatonic) = 120

Chatuswara Kuta Tana (Tetratonic) = 24

Triswara Kuta Tana (Tritonic) = 6

Dwiswara Kuta Tana (Bitonic) = 2

Monotone = 1

(37-38)

5. एकस्वरादि-कूटतानचतुष्टयस्य नामानि

आर्चिको गाथिकश्चाथ सामिको ऽथ स्वरान्तरः ।
एकस्वरादितानानां चतुर्णामभिधा इमाः ॥३९॥

We know the names of Kuta Tana with six and five notes as Shadav and Audav respectively. Now we

249

define the names of Kuta Tana with 4, 3, 2 and 1 notes respectively.

'Swarantara' is the name of Kuta Tana with 4 notes. (Tetratonic)

'Samika' is the name of Kuta Tana with 3 notes. (Tritonic)

Gathika' is the name of Kuta Tana with 2 notes. (Bitonic)

Archika' is the name of Kuta Tana with 1 note. (Monotonic)

(39)

'Archika' – This means literally related to Rigveda. In Rig Veda, the sacred hymns were sung in a monotone and hence the name. The monotones were of three formats; Uddat (raised) Anuddat (lower) and Swarit (intermediate or medium pitch).

'Gathika' –Bi-tonal structure related to 'Gatha', which are verses conveying non-religious content.

'Samika' - Tri-tonal structures for religious hymns were used in Sama Veda. The Tritonal hymns of Sama Veda are called Sama Gana and hence the name 'Samika'.

Shadav Kuta Taan (Hexatonic)

6. पाडवकूटतान-सङ्ख्या

उक्ता: शुद्धादिभेदेन निगयुक्ताश्चतुर्विधाः ।
तयोरेकंकहीनास्तु द्वेधा मूलक्रमा मताः ॥४०॥
षड्जाद्यौ मध्यमाद्यौ च चत्वारः स्युर्द्विधा द्विधा ।
चतुर्धा ऽन्ये दशेत्यष्टाचत्वारिंशदमी क्रमाः ॥४१॥
सर्विंशतिः सप्तशतो प्रागुक्ता गुणिता क्रमैः ।
चतुर्स्त्रिंशतसहस्राणि षष्ट्या पञ्च शतानि च ॥४२॥
इति षाडवसङ्ख्या स्यात्,

720 Hexatonic 'Shadav' Kuta Taans can exist for EACH Murchana. However to create Kuta Taan the series of swara has to be omitting one-note and while doing so 48 murchanas can be counted. Thus, the total number of Hexatonic permutational note-series;

 'Shadav Kuta Tana' are 720 Shadav permutations x 48 murchanas = 34,560 (40-42)

Audav Kuta Taan (Pentatonic)

7. औडुवकूटतान-सङ्ख्या

,अथ पञ्चस्वरान्ब्रुवे ।
गाद्यौ धाद्यौ निषादाद्यौ चतुर्भेदाः षडौडुवाः ॥४३॥
अष्टावन्ये द्विचेत्येवं चत्वारिंशदिमे क्रमाः ।
सर्विंशतौ शते तेंश्च गुणिते ऽष्टौ शतानि तु ॥४४॥
चत्वारि च सहस्राणि सङ्ख्या पञ्चस्वरेष्विति ।

120 Pentatonic 'Audav' Kuta Taans can exist for EACH Murchana. However to create Audav Kuta Taan the

251

series of swara has to omit the last TWO notes of the murchana-krama series, and while doing so 40 murchanas can be counted. Thus, the total number of pentatonic permutational note-series;

'Audav Kuta Tana' are 120 Audav permutations x 40 murchanas = 4,800 (43-44)

Total Kuta Taan

As we calculated the Shadav and Audav Kuta Taan numbers, we arrived through the analysis of Murchanas and Kuta Taan permutations following TOTAL summary of all Kuta Taan that can exist.

8. चतुःस्वरकूटतान-सङ्ख्या

चतुःस्वरेषु न्याद्यौ द्वौ चतुर्धा द्वादशापरे ॥४५॥
क्रमा द्विधेति द्वात्रिशच्चतुर्विशतिताडिता ।
शतानि सप्ताष्टषष्टच्या स्याच्चतुःस्वरसंमितिः ॥४६॥

9. त्रिस्वरकूटतान-सङ्ख्या

त्रिस्वरेषु तु माद्यौ द्वावमेदौ वादशापरे ।
द्विधा षड्विंशतिरिति क्रमास्ते षड्भिराहताः ॥४७॥
षट्पञ्चाशच्छतं च स्युः,

10. द्विस्वरैकस्वर-कूटतान-सङ्ख्या

, द्विस्वरेषु पुनर्द्विधा ।
रिगधन्याद्वयो ऽष्टौ स्युः शुद्धास्तदितरे क्रमाः ॥४८॥

द्वार्विशतिस्ते तु चतुश्चत्वारिशद् द्विताडिताः ।
एकस्वरास्त्वमेदत्वान्मौला एव चतुर्दश ॥४९॥

Number of Notes	Kuta Taan Type	Kuta Series Name Sanskrit	Kuta Series Name English	P Permu- tations	M Murchana Series	TOTAL= PxM
7	Complete	Purna	Complete	5040	56	282,240
6	Incomplete	Shadav	Hexatonic	720	48	34,560
5	Incomplete	Audav	Pentatonic	120	40	4,800
4	Incomplete	Chatuswara	Tetratonic	24	32	768
3	Incomplete	Triswara	Tritonic	6	26	156
2	Incomplete	Dwiswara	Bitonic	2	22	44
1	Incomplete	Ekswara	Monotonic	x	14	14
		Total Kuta Taans including original series				**322,582**

Note: The 322,582 Taans include many permutations that are repetitive between Shadaj and Madhyam Grama. On detailed analysis of these duplicative Taans, it is observed that there are 4,652 duplicative Kuta Taans. These are broken down as follows;

1. Complete Series (Mula Murchana Kramas) – 56x7 = 392 Duplicative Taans

2. Incomplete Series - Total 179 Duplicative
    a) Hexatonic = 48 Duplicative
    b) Pentatonic = 40 Duplicative
    c) Tetratonic = 32 Duplicative
    d) Tritonic = 26 Duplicative
    e) Bitonic = 22 Duplicative
    f) Monotonic = 11 Duplicative

3. Repeated Taans – Total 4081
    a) Shuddha Madhya Murchana of Madhyam Grama – 63
    b) Margi Murchana of Madhyam Grama – 593
    c) Pauravi Murchana of Madhyam Grama – 3425

These Three add up to 392+179+4081 = 4652 Repetitions

Total Kuta Taans including original series		*322,582*
Total Kuta Taan Repetitions	(-)	4652
Exact number of Unique Kuta Taan Series		**317,930**

Kuta Taan Prastara

(v) कूटतान-प्रस्तारः

क्रमं न्यस्य स्वरः स्थाप्यः पूर्वः पूर्वः पराद्धः ।
स ॰ चेदुपरि तत्पूर्वः पुरस्तूपरिवर्तिनः ॥६२॥
मूलक्रमक्रमात्पृष्ठे शेषाः प्रस्तार ईदृशः ।

We now study the 'Prastara' i.e. the methodology of formation of the Kuta Taans.

i. Identify which type of Kuta Taan permutations one needs to expand upon. I.e. Complete (including 7 notes), Hexatonic, Pentatonic, Tetratonic, Tritonic, or Bitonic.

ii. Start by establishing and identifying the original natural order of seven notes s r g m p d n (if complete series) OR s r g m p d (for incomplete Hexatonic series) and so on…till we establish the original order of monotonic series as per our needs of the Kuta Taan. ***This forms the first natural note series.***

iii. Starting from left most note (preceding note), initiate creating the permutational series by placing preceding note below the succeeding one. Notes following the succeeding note are placed after (to the right) the transposed note. Notes preceding the succeeding note are placed

254

before (to the left) of the succeeding note in their *original natural order*. ;

iv. If the preceding note of the first step becomes the succeeding note of next step then preceding note may be advanced by one keeping subsequent notes afterward;

v. The rest of the notes in series are left behind (to the left of) the placement of the succeeding note per the original order of notes. The preceding note of the first series becomes a succeeding note of the second series and so on.

vi. This process is repeated until all permutational series have been listed for expansion.

This is the systematic methodology of 'expanding' and applying numerical permutational calculus to arrive at Kuta Taans. Since the objective is to identify the forms of unique note-series, no relationship can be repeated as well as a set of transposed notes in a given series can be interchanged only once. (62)

We earlier observed that complete or incomplete murchanas with their notes produced in permutation format (non-sequential) create permutational note series known as 'Kuta Taan'. Here we now study the most important practical process of creating the Kuta Taan. We analyze the permutational mathematical methodology to derive the possible 317,930 unique Kuta Taans.

This process is involved and can be understood with a specific example. As such, we will take our time to understand the application of numerical science of permutations to recreate forms of the Kuta Taans in a given series. This is the process of placing notes of various types of note-series (Kuta Taan) such as complete, hexatonic, pentatonic, tetratonic, etc. in a specific progressive sequence of formation of notes giving rise to all possible permutations of that particular note-series in consideration.

Tetratonic Kuta Taan Formation Example

24 Kuta Taan series can be formed with 4 notes

Note - Series Sr. #	Notes Transposed	Series	Relative Numerical Order	Methodology Process
1st Taan	None	srgm	1-2-3-4	Natural order of first series identified
2nd Taan	s for r	rsgm	2-1-3-4	To form the SECOND series the preceding note = 's' is written below the succeeding note = 'r'. Notes following the succeeding note 'r' here are 'g m' and they are placed to the right of the transposed preceding note as in original series. Notes prior to the succeeding note are placed to the left of succeeding note in their original order. Here there are no notes before succeeding note 'r' besides the original 's' which is transposed as per the initial need. Thus forming second series r s g m.
3rd Taan	r for g	sgrm	1-3-2-4	To form the THIRD series we focus on second series as its starting series. Preceding note = 'r'. Per rule write it below 's' immediately succeeding note, but this produces s r g m again which is a duplicate of the first Taan. So, 'r' and 's' can't be transposed as that relationship has already been used. So we continue to move to right to transpose 'r' with succeeding note 'g'. Write 'r' below 'g' in the new series. Now notes to right of 'g' are 'm' and it is placed to right of the transposed 'r'. Notes to left of 'g' are 's' and they remain in their original order before the succeeding note 'g'. Thus forming third series s g r m.
4th Taan	s for g	gsrm	3-1-2-4	To form the FOURTH series we focus on third series as its starting series. Preceding note = 's'. Per rule write it below 'g' immediately succeeding note. Now notes to right of 'g' are 'r m' and are placed to right of the transposed 's'. Here there are no notes before succeeding note 'g' besides the original 's' which is transposed as per the initial need. Thus forming fourth series g s r m.
5th Taan	g for r	rgsm	2-3-1-4	To form the FIFTH series we focus on fourth series as its starting series. Preceding note = 'g'. Per rule write it below 's' immediately succeeding note, but this produces s g r m again which is a duplicate. So, 'g' and 's' can't be transposed as that relationship has already been used. So we continue to move to right to transpose 'g' with succeeding note 'r'. This produces r s g m which is also a duplicate. So we conclude that 'g' can't be transposed and so next possible preceding note now becomes 's' and succeeding note to be transposed is 'r'. So we transpose them and notes to right of succeeding note 'r' are 'm' which stay as it is. Notes to left of 'r' are 'g s' and they retain their natural order before the transposed note. Thus forming fifth series r g s m.
6th Taan	r for g	grsm	3-2-1-4	To form the SIXTH series we focus on fifth series as its starting series. Preceding note = 'r'. Per rule write it below 'g' immediately succeeding note. Write 'r' below 'g' in the new series. Now notes to right of 'g' are 's m' and it is placed to right of the transposed 'r'. Notes to left of 'g' are 's' and they remain in their original order before the succeeding note 'g'. Here there are no notes before succeeding note 'g' besides the original 'r' which is transposed as per the initial need. Thus forming sixth series g r s m.
7th Taan	g for m	srmg	1-2-4-3	To form the SEVENTH series we focus on sixth series as its starting series. Preceding note = 'g'. Per rule write it below 'r' immediately succeeding note, but this produces g s m again which is a duplicate. So we continue to move to right to transpose 'g' with succeeding note 's', but this produces r s g m again which is a duplicate. So preceding note 'g' must be now transposed with last option 'm' as succeeding note. There are no notes after 'm', so that does not need to be factored in. There are notes before succeeding note 'm' and they are 'r s'. They will be written in their original natural order as 's r' to the left of the succeeding note 'm'. Thus forming seventh series s r m g.

Similar process must be used to arrive at all 24 Taans of Tetratonic permutational series.

It is interesting to note here that Sage Bharat in his Natya Shastra does not allude to 'Kuta Taan' and Sharangdev as a result limits this concept only to Shadaj and Madhyam Grams which are considered to be 'Deshi' in their musical qualities. 'Deshi' literally means regional and music for the common person as opposed to the music that is performed by divine and semi-divine beings.

Gandhar Gram is the third Grama that we discussed early on, used by the divine beings. This is known as 'Margiya' music. It is natural to assume that incomplete (liberal interpretations) Taans would not be permitted in the divine format of music for the divinities and hence only 'Shudha' pure Taans exist there.

Merukhand

(vi) खएडमेरु:

1. खण्डमेरु-निर्माणम्

सप्ताद्येकान्तकोष्ठानामधो ऽध: सप्त पङ्क्तय: ॥६३॥
तास्वाद्यायामाद्यकोष्ठे लिखेदेकं परेषु खम् ।
वेद्यतानस्वरमितान्यस्ये त्तेष्वेव लोष्टकान् ॥६४॥
प्राक्पङ्क्तयन्त्याङ्कसंयोगमूर्ध्वाध:स्थितपङ्क्तिषु ।
शुभ्यादधो लिखेदेकं तं चाधो ऽध: स्वकोष्ठकान् ॥६५॥
कोष्ठसङ्ख्यागुणं न्यस्येत्खण्डमेरुरयं मत: ।

स	रि	ग	म	प	ध	नि
१	०	०	०	०	०	०
	१	२	६	२४	१२०	७२०
		४	१२	४८	२४०	१४४०
			१८	७२	३६०	२१६०
				९६	४८०	२८८०
					६००	३६००
						४३२०

The total numbers of permutations are 317,930 as derived earlier. Now a table known as the 'Merukhand' table is presented which assists in the permutational analysis if one wanted to approach it from a purely mathematical point of view. This relates to two techniques that are named 'Nashta' and 'Udishta' respectively. (22-26)

The numbers in the squares of the 'Merukhand' table represent the permutational calculus derived from the permutational note series discussed in the 'Kuta Taan' section. 'Nashta' is the name of a musical problem where one is given a precise numerical number (out of the 317,930) of the note-series or 'Kuta Taan' and the objective is to identify its tonal format giving its notes and their order. 'Uddishta' is the opposite musical problem where one knows the tonal format of the note-series but wants to identify its exact numerical serial number. Thus, these two musical problems are for those mathematically curious seekers who want to analyze the 'Kuta Taan' and identify their serial numbers and vice versa. Since this is purely a theoretical and mathematical process exercise, we will limit our discussion to this extent.

4. खण्डमेरुत एकस्वरादितानानां सङ्ख्या-परिज्ञानोपायः
तानस्वरमितोर्ध्वाधःपङ्क्तिगान्त्याङ्कमिश्रणात् ।
एकस्वरादितानानां संख्या संजायते क्रमात् ॥७१॥

Besides the Nashta-Uddishta process, the Khandmeru also assists in quantifying the total number of note-series in each of the seven types of permutational 'Kuta Taan' i.e. Hexatonic, Pentatonic, all the way down to Monotonic. However, the most important aspect of the

Khand Meru is that it provides complete proof of the mathematical and permutational foundation of the musical order discussed so far.

For those interested in going deeper into mathematics alone, it is suggested to refer to much widely available documentation on 'Sangeet Ratnakar' and Khand Meru that describe the Nashta-Uddishta methodology of numerical proof. From a practical singing point of view, the Nashta Uddishta mathematical process on its own does not help us much with the quality of singing. It does help rather solely in identifying the logical relationships between the note-series and their serial numerical order amongst the 31,930 unique Kuta Taans.

Section 4 – Sadharana, Varnalankara, Jati, Giti

Thus far, we have covered the first six pranas of the 'Melody' in the Indian musical system. The elements of Nada, Shruti, Swara, Grama, Murchana, and Taan have given us the foundation of now applying those concepts in practical terms. In this section, we discuss the 'implementation' ideas that evolve from the concepts that have been discussed as they relate to the 'Swara Shashtra' so far.

Sadharana

4 Sadharana (Overlapping)

अथ पञ्चमं साधारणप्रकरणम्

(i) द्विविधं साधारणं, तत्र स्वरसाधारणम्

साधारणं भवेद् द्वेधा स्वरजातिविशेषणात् ।
स्वरसाधारणं तत्र चतुर्धा परिकीर्तितम् ॥१॥

काकल्यन्तरषड्जेश्च मध्यमेन विशेषणात् ।

'Sadharana' is the concept of overlapping tones. It is of two types as it pertains to 'Swara' (notes) and as it pertains to 'Jati' (melodic types). Of these, the Swara based overlapping concept in practice is of four types. i.e.

I.   Kakali Sadharana

II.   Antara Gandhar Sadharana

III.   Shadaj Sadharana &

IV.   Madhyam Sadharana (1)

In normal Sanskrit usage, the word 'Sadharana' means literally something basic or of normal quality. However, in the context of 'Swara Shashtra', this word has attained a unique context and it means 'overlapping tonality' of melody. How did this happen?   To understand this we need to trace the sensibilities and emotional 'bhavatmak' genesis of this term in 'Swara Shastra'.

The musical Sadharana concept has been mentioned in the earliest references in sage Bharat's 'Natya Shastra' treatise. Bharat and other founding sages of the Indian musical system were living in complete harmony with nature around them. As such, we see repeatedly in the "Swara Shastra' (Science of Melody) as well as 'Taal Shashtra' (Science of Rhythm) the natural order of the universe and the seasons being intertwined in the creation and practice of musical and rhythmic concepts. This present idea is one more proof of such influence of nature in the evolution of music.

In Sanskrit, the term "Kaal Sadharana" signifies time and its seasonality and the transitional periods between seasons. This is very evident for example in four seasonal concepts of the western hemisphere where there is spring, summer, fall, and winter. In the Indian context, these seasons are named Vasant Kaal, Ushna Kaal, Sharad Kaal, and Shit Kaal respectively. During

261

transitions between the seasons, there is a period of overlap. This period has qualities of both seasons on its border and creates a unique element of experience on its own creating a new dimension.

This can be explained by the physical experience of venturing out in times in winter when the Sun is very harsh and bright and one can get a sunburn by being exposed to it yet the body feels cold when one-steps outside. Thus, one experiences summer, which has not arrived yet winter has not ended. The feeling is an experience of an overlapping sense of the seasons. This idea denotes the overlap of the end portion of the prior situation and the beginning portion of a new situation creating a completely new situation that partakes in experience of qualities of both prior and the subsequent situations (Harsh Sun Rays and Cold Air of winter). This new experience is different from both original situations and is the 'Sadharana' in essence.

Twofold overlapping note intervals

In melodic note terms, Bharat defines this as a transition phase between two notes where the tonal quality of one-note 'rises' and 'overlaps' between the prior and subsequent notes. We discuss now the note overlapping or the Swara Sadharana four types next.

There are four notes that are relevant for the Swara Sadharana and they are Kakali (Nishad), Antara (Gandhar), Shadaj and Madhyam. In the process of transition between these notes, the Shruti tonality can be overlapped between the prior and subsequent notes in question producing the four types of Swara Sadharana named in the verse above. We shall discuss these in detail now.

Overlapping of Kakali and Antara

(ii) काकल्यन्तरसाधारणम्

साधारणः काकली हि भवेत्षड्जनिषादयोः ॥२॥

साधारण्यमतस्तस्य   यत्तत्साधारणं   विदुः ।
अन्तरस्यापि   गमयोरेवं   साधारणं   मतम् ॥३॥

Overlapping 'Sadharana' of Kakali and Antara. The name of the Nishad that produces this essence of overlapping emotion between tonality of Nishad and Shadaj is called Kakali. Hence, this is also called Kakali Sadharana. In a similar manner, overlapping Antara occurs between Gandhar and Madhyam. (2-3)

'Kakali Sadharana' – In the overlapping process 'Sadharana' of Kakali 2 notes Nishad and Shadaj are used. The 'Kakalization' of Nishad occurs by overlapping the shrutis from Shadaj onto Nishad. In standard Shadaj Grama definition, Nishad has two Shrutis numbers 21 and 22 assigned to it and Shadaj has four Shrutis numbers 1, 2 3 and 4 assigned to it. Kakalization occurs when Nishad gains Shruti numbers 1 and 2 of Shadaj subsequent to its original Shruti numbers 21 and 22. Thus, 'Kakali Nishad' is born containing characteristics of both Nishad and Shadaj and is four Shrutis long with its final note Shruti occurring on 2nd Shruti. (In pure sequential order 24th, but since there are only 22 Shrutis that are considered the Shruti cycle reinitiates at the higher register after 22nd and hence the ending note Shruti of Kakali is on 2nd Shruti of the next higher register.)

The question arises as to why is it named Kakali Nishad when it affects both Shadaj and Nishad. The answer is that 'Kakali' absorbs complete Nishad and gains its full character but from Shadaj it only gains partial character leaving note Shruti (4ᵗʰ) of Shadaj intact.

'Antara Sadharana' – In the overlapping process 'Sadharana' of Antara Gandhar 2 notes Gandhar and Madhyam are used. The 'Antarization' of Gandhar occurs by overlapping two shrutis from Madhyam onto Gandhar. In standard, Shadaj Grama definition Gandhar has two Shrutis numbers 8 and 9 assigned to it and Madhyam has four Shrutis numbers 10, 11, 12 and 13 assigned to it. 'Antara Sadharan' occurs when Gandhar gains Shruti numbers 10 and 11 from Madhyam subsequent to its original Shruti numbers 8 and 9. Thus, 'Antara Gandhar' is born containing characteristics of both Gandhar and Madhyam and is four Shrutis long with its final note Shruti occurring on 11th Shruti.

Usage of Kakali and Antara

(iii) काकलयन्तर-प्रयोगविधिः

प्रयोज्यौ षड्जमुच्चार्य काकलीधेवतौ क्रमात् ।
एवं मध्यमुच्चार्य प्रयुङ्जीतान्तरर्षभौ ॥४॥
षड्जकाकलिनौ यद्घोच्चार्य षड्जं पुनर्व्रजेत् ।
तत्परान्यतमं चैवं मध्यमं चान्तरस्वरम् ॥५॥
प्रयुज्य मध्यमो ग्राह्यस्तत्परान्यतमो ऽथ वा ।
अल्पप्रयोगः सर्वत्र काकली चान्तरः स्वरः ॥६॥

'Having defined the precise manner of Kakali and Antara notes, we discuss how to deploy them in practical usage now. The usage of both these notes is done in mostly the 'Descending' (Avroh) sequence of pronunciations. So after singing S, Kakali followed by

D is employed. Similarly, after singing M, Antara followed by R is employed.

Another way of usage is the pairing S-Kakali-S and then move on to any succeeding note R, G, M, P or D.

Similarly, for Antara, usage is M-Antara Gandhar-M followed by any of the succeeding notes P, D, N, S or R.

These Kakali and Antara notes are rarely used in standard singing. (4-6)

Overlapping of Shadaj and Madhyam

(iv) षड्जमध्यमसाधारणम्

निषादो यदि षड्जस्य श्रुतिमाद्यां समाश्रयेत् ।
ऋषभस्त्वन्तिमां प्रोक्तं षड्जसाधारणं तदा ॥७॥
मध्यमस्यापि गपयोरेवं साधारणं मतम् ।
साधारणं मध्यमस्य मध्यमग्रामगं ध्रुवम् ॥८॥
साधारणे कैशिके ते केशाग्रवदणुत्वतः ।
ते एव कैश्चिदुच्येते ग्रामसाधारणे बुधैः ॥९॥

Overlapping 'Sadharana' of Shadaj and Madhyam. For 'Shadaj Sadharana' – Nishad takes the first Shruti from Shadaj and Rishabh takes the last Shruti of Shadaj. Thus, Shadaj remains of two Shrutis numbers 2 and 3.

Similarly, for 'Madhyam Sadharana' Gandhar takes on the first Shruti of Madhyam and Pancham takes on the last Shruti of Madhyam. Thus, Madhyam becomes shorter of two Shrutis numbered 11 and 12. Both the overlapping concepts are extremely fine in their emotions and hence the name 'Kaishika' fine like the point of hair 'Kesh' is given to them. (7-9)

We must note here that the longest Shruti interval that

can exist in the Grama concept is a note of four Shrutis. Recall that the difference between Shadaj and Madhyam grama is that Pancham in Shadaj Grama is of 4-Shruti interval and in Madhyam Grama, it is of 3 Shruti interval. Thus if we deploy the Madhyam Sadharan in Shadaj Grama its Pancham will gain one Shruti from Madhyam and swell to five Shrutis which is not permitted or rather not ideal in practice. Thus for most experts, Madhyam Sadharana is limited to Madhyam Grama only.

Overlapping of Jatis

(v) जातिसाधारणम्

एकग्रामोद्भुवास्त्वेकांशासु   जातिषु   यद्भुवेत् ।
समानं   गानमार्यास्तज्ज्जातिसाधारणं   जगुः ॥१०॥
जातिसाधारणं   केचिद्रागानेब   प्रचक्षते ।

The concept of 'Jati' will be introduced later in the book. Here in the context of Sadharana, one must understand that the identical melodic sequences amongst Jatis of the same Grama (having the same fundamental note) are the overlapped melodic constructs. Some identify the Jati Sadharana concept with 'Ragas'.

Author's Instrument of Laya Experimentation at Krishna
Leela, Tulsi Ghat, and Varanasi, India

Varnalankar

ब्रथ षष्ठं वर्णालङ्कारप्रकरणम्
(i) वर्णलक्षणं, तद्भेदाश्च

गानक्रियोच्यते वर्णः स चतुर्धा निरूपितः ।
स्थाय्यारोह्यवरोही च संचारीत्यथ लक्षणम् ॥१॥
स्थित्वा स्थित्वा प्रयोगः स्यादेकस्यैव स्वरस्य यः ।
स्थायी वर्णः स विज्ञेयः परावन्वर्थनामकौ ॥२॥
एतत्संमिश्रणाढर्णः संचारी परिकीर्तितः ।

4 Varna (Tone Patterns)

Now, we take another step towards further enhancing our understanding of the 'practical' aspects of applying the science of melody in our musical process. The concept of 'Varna' is introduced here.

During the process of voice-production and act of singing, the word 'Varna' represents the tonal pattern. This is of fourfold classification.

1. Sthayi (steady and halting tonal pattern such as, 's s s', 'r r r', etc.),

2. Arohi, (ascending tonal pattern such as 'srgmpdn'),

3. Avrohi (descending tonal pattern 'ndpmgrs') and

4. Sanchari (circular tonal pattern. Combination of all of the above with less emphasis on sequential

order but rather circular in nature such as 's s s n pd pd m n s' etc.) (1-2)

The Sanskrit word 'Varna' has many meanings such as syllable, color, taste, social order, etc. Here in the context of music, we mean it in the sense of 'order of tonal value of the melody'. This tonal order then provides melodic color and special emotional sense to the music and that is the essence of musical 'Varna'.

This concept is essentially most fundamental in Vedic linguistic science. The 'Varna' represents a Syllable. The syllables then go onto form the words that convey some idea. The words when put together form a sentence that conveys the collective concept of the words. Sentences when collected in our language in the form of prose or poetry then culminate to convey the full story of the entire thought process behind those words and their 'Varna'. Similarly, in a musical sense, the Shrutis (tones) organized into a specific order form special melodic phrases (tonal patterns - Varna). These Varna then become units of melodic compositions and form musical prose and poetry such as a 'Bandish' literally meaning a melodic and poetic idea 'enclosed' in a group of words.

The fourfold classification of 'Varna' is described in the oldest known musical treatise from sage Bharat. These tonal patterns are used in a specific order and are not used in isolation to create melodic compositions. The predominant characteristic of the melodic phrase determines its 'Varna' and there is a specific order in different styles of music as to how the Varnas are deployed. If the melodic phrase is 'Steady' in nature by staying at one tone for a while and 'harping' on it then it is called 'Sthayi'. The Arohi and Avrohi are self-evident from their definitions. What needs further elaboration is the idea of 'Sanchari'.

Sage Matanga, in his musical treatise 'Brihaddeshi' defines 'Sanchari Varna as that pattern where tones move in a note-series (Taan) one by one or two at a time or are joined together joining the respective final notes (with initial notes of the successive units). E.g.' s s s n m m n m p n r r p p n p n p n d'

Alankara (Tonal Embellishments)

(ii) अलङ्कारलचर्णं, तद्भेदानिरूपणं च

(क) अलङ्कारलक्षणम्

विशिष्टं   वर्णसन्दर्भमलङ्कारं   प्रचक्षते ।।३।।

'Alankara' in musical parlance means specifically 'a unique and specific, stylistic melodic phrase progression'. Just as a jeweled ornament embellishes the beauty of its beholder, in the same way, 'Alankara' (literal meaning- ornament) are tonal embellishments created by using various 'jewel' like tonal patterns (Varna) in their construction. What is being beautified by the usage of 'Alankara' is the Melodic composition being produced by the practitioner of the music. (3)

Tonal embellishments are extremely important ingredient of melodious music however, their usage must be employed in a certain order and regulations of the tonal patterns such that the patterns of tone 'Varna' maintain their unique character at all times. Alankara are thus combinations of various melodic movements. It is the "Varna' that make up the 'Alankara' and hence we derive the precise sensibility of 'Alankara' (tonal embellishments) as "the particular and unique manner of deploying the usage of the four 'Varna' (tonal pattern types) in the creation of a melodic rendering". The uniqueness of deploying the tone pattern is based on the specific style of singing and the 'Kala' (artistic prowess) of the performer.

Sage Bharata illustrates the sensibility of 'Alankara' as follows; "Melody without ornamentation will be like a night without moon, a river without water, a creeper without its flowers and a lady without ornaments."

"Melody must be embellished by these 'ornamentations' without damaging the true nature of the tonal patterns 'Varna'. For ornaments must be used and put in their proper places on the body and the wrong ornament in the wrong place would create disharmony."

Sthayi Varana (Steady tone pattern)

(ख) स्थायिवर्णगतालंकारोद्देशः

तस्य भेदास्तु बहवस्तत्र स्थायिगतान्बुवे ।
येषामाद्यन्तयोरेकः स्वरस्ते स्थायिवर्णगाः ॥४॥
प्रसन्नादिः प्रसन्नान्तः प्रसन्नाद्यन्तसंज्ञकः ।
ततः प्रसन्नमध्यः स्यात्पञ्चमः क्रमरेचितः ॥५॥
प्रस्तारो ऽथ प्रसादः स्यात्सप्तैते स्थायिनि स्थिताः ।

Theoretically speaking there are infinite varieties of Alankara. (The limitation to quantify the numerical and melodic 'form' is not a limitation of music but rather that of human intellect' – author). However, during Sharangdev's time, 63 Alankara were in use and hence those are described subsequently as examples.

Here the 'Sthayi' steady variety is described. The Alankara (tonal patterns) that have the same Shruti (tone) in start and end, though tonal repetitions occur in between, are said to be of Sthayi Varna. There are seven sub-categories of 'Sthayi' Alankara as defined by Sharangdev. Viz.

    I.   Prasannadi (Mandra tone in beginning)

    II.   Prassananta (Mandra in end)

    III.   Prasannadyanta (Mandra in start and end)

IV.   Prasannamadhya (Mandra in middle)

V.   Kramrechita (devoid of order)

VI.   Prastara (extension)

VII.   Prasada (clarity)

Prasanna is another name for Mandra Shruti. All these seven types will be described with examples now. (4-5)

Mandra, Tara, Madhyam, etc.

(ग) मन्द्रतारलचचणाम्

मन्द्रः प्रकरणे ऽत्र स्यान्मूर्च्छनाप्रथमः स्वरः ॥६॥

स एव द्विगुणस्तारः पूर्वः पूर्वो ऽथ वा भवेत् ।

मन्द्रः परस्ततस्तारः प्रसन्नो मृदुरित्यपि ॥७॥

मन्द्रस्तारस्तु दीप्तः स्यान्मन्द्रो बिन्दुशिरा भवेत् ।

ऊर्ध्वरेखाशिरस्तारो लिपौ त्रिर्वचनात्प्लुतः ॥८॥

We introduced the term Mandra in the previous verse. Now let us define Mandra and its related terms Tara and Madhyam in the context of tonality.

'Mandra' (lower pitch) signifies the first tone of the Murchana. When this Shruti (tone) is doubled in intensity, it becomes 'Tara' (higher pitch) this corresponds to the eighth Swara (note). This is nothing but the tonality shifting from one register to the other higher one. Thus, Mandra is the Shruti of the previous register to the Tara (which has a double intensity of Shruti than Mandra). So traversing between the three registers these two terms Mandra and Tara are relative concepts with a double pitch relationship respectively.

Mandra is also known as Prasanna or Mridu while Tara's tone is also known as 'Dipta'. The intensity relationship in terms of Sanskrit grammatical rules is akin to the degree of the length of vowels of the pronunciations.

Mandra = (grammatical Hrsva short vowel length of one 'Matra' degree) denoted by writing a dot (.) on top of the note.

Tara = Double of Mandra (grammatical Deergha longer length = 2 Matra.) Denoted by writing a vertical dash (') on top of the note.

Taratara = Triple intensity of Mandra = (grammatical Pluta = triple length = 3 Matra.) Denoted by writing a superscript (₃)on top of the note. (6-8)

We must note here that this "Mandra – Tara" is a purely relative concept and as often mistakenly equated with three registers. In essence if the 'prior' tone is in the lower register (in the triple register model) than it is called Mandra and the Tara will be in the middle register of double intensity. Alternatively, if the 'prior' tone is in the middle register than that would be Mandra and the subsequent Tara relative to it will be in the highest (third) register. So Mandra and Tara do not define registers they just define the relative aspect of their tonal relationship. Madhyam is essentially in between Mandra and Tara pitches.

Sage Bharata conveys the same idea but emphasizes in his treatise the locations of sound production in the human body as we discussed in our Yogic analysis of Naad. Therefore, besides the tonal relationship of Mandra and Tara, the location from where these sounds are produced (viz. lowest below chest, chest, heart, and head and highest near cerebral cortex) coupled with their grammatical intonation value ('Kaku' bheda) determines the excellence of the verbal essence of the melodic content (Pathya Guna).

7 Sthayi Alankara

The table below provides clarity with examples of the seven types of Sthayi Alankaras. The first Murchana of the Shadaj Grama whose name is UttaraMandra is used here for creating examples. The first four Sthayi Alankara numbers 1 to 4 and then the last three numbers 5 to 7 form two distinct groups with a common identity. The first group has the same note in beginning, middle and ending with the difference among placement of higher and lower registers. The second group has a beginning and ending notes the same still differing in higher and lower notes amongst registers, but also the middle portion is different in the second group.

Sthayi Alankara

			English notes without higher lower symbols.
1	Prasannadi	Mandra in beginning - Two lower tones followed by one higher tone. Start middle and end all are same notes.	s s s
2	Prassananta	Mandra at end - One higher tone followed by two lower tones. Start middle and end all are same notes.	s s s
3	Prasannadyanta	Mandra in start and end - Two lower tones with one Higher tone in middle. Start middle and end all are same notes.	s s s
4	Prasannamadhya	Mandra in middle - Lower tone in middle between tow higher tones. Start middle and end all are same notes.	s s s
5	Kramrechita	This is a three 'Kala' or three part Alankara. The tonal value of the notes in between the beginning and end of the phrases doubles and then triples. The pitch intensity is 1-2-3 relationship in the middle sections of the three parts. Start and end notes are same middle notes are NOT same.	s r s s g m s s p d n s
6	Prastara	Extension. This 'extends' Kramarechita. While Start and end notes are same middle notes are NOT same UNLIKE kramarechita the ending tone is always Tara here.	s r s s g m s s p d n s
7	Prasada	Clarity. This is reverse of Prastara.	s r s s g m s s p d n s

12 Arohi-Avrohi (Ascendant/Descendent)

(ङ) आरोहिवर्णगतालंकारोद्देशः

स्यातां विस्तीर्णनिष्कषौ बिन्दुरम्पुच्चयः परः ।

हसितप्रेङ्गिताक्षिप्तसन्धिप्रच्छादनास्तथा ॥१४॥

उद्गीतोद्वाहितौ तद्वत्त्रिवर्णा वेणिरित्यमी ।

द्वादशारोहिवर्णस्थालङ्काराः परिकीर्तिताः ॥१५॥

	Arohi Avrohi Alankara	The examples used are Arohi here. If these are reversed in Descending order then it becomes Avrohi Alankara	
			English notes without higher lower symbols.
1	Vistirna	Ascent takes place from first note of Murchana to successive notes with tone elongated twice the length of a normal single tone interval. After the doubly elongated note one proceeds to the next higher note.	s r g m p d n
2	Nishkarsha	Same as Vistirna but here 'Short' tone notes are deployed twice, thrice or four times in same space.	s-s r-r g-g m-m p-p d-d n-n
			s-s-s r-r-r g-g-g m-m-m p-p-p d-d-d n-n-n
			s-s-s-s r-r-r-r g-g-g-g m-m-m-m p-p-p-p d-d-d-d n-n-n-n
3	Bindu	Here Ascending tones are deployed in four groups of this order; triple {prolonged (pluta), short (Hrsva)} x3 prolonged.	s3r g3m p3d n3
4	Abhyuchaya	Ascending notes dropping alternate notes.	s g p n
5	Hasita	Here tones ascend with increased number of repetitions as the corresponding note in sequence.	s r-r g-g-g m-m-m-m p-p-p-p-p d-d-d-d-d-d n-n-n-n-n-n-n
6	Prenkhita	Here the first two notes are sung and then on singing the third its is preceded by second by moving one step back...and thus it repeats.	s-r r-g g-m m-p p-d d-n n-s
7	Akshipta	Here first pair is formed by dropping the second note so it is 1-3, 3-5, 5-7	s-g g-p p-n
8	Sandhipracchadan	Here the first phrase (kala) is of three tones and there are two more phrases the relationship is 1-2-3, 3-4-5, 5-6-7	s-r-g g-m-p p-d-n
9	Udgita	Here there are two phrases of three tones each with relationship as 1-1-1-2-3, 4-4-4-5-6	s-s-s-r-g m-m-m-p-d
10	Udvahita	Here there are two phrases of three tones each with relationship as 1-2-2-2-3, 4-5-5-5-6	s-r-r-r-g m-p-p-p-d
11	Trivarna	Here there are two phrases of three tones each with relationship as 1-2-3-3-3, 4-5-6-6-6	s-r-g-g-g m-p-d-d-d
12	Prithagveni	Here there are two phrases of three tones each with relationship where each note is repeated thrice as 1-1-1, 2-2-2, 3-3-3, 4-4-4, 5-5-5, 6-6-6. Note this is similar to one of the Nishkarsha Alankara variety except that here there is NO n note used as only first six notes are used.	s-s-s r-r-r g-g-g m-m-m p-p-p d-d-d

26 Sanchari Alankara (Circular Tone patterns)

(छ) सञ्चारिवर्णगतालंकारोद्देशः

मन्द्रादिर्मन्द्रमध्यश्च मन्द्रान्तः स्यादतः परम् ॥२६॥
प्रस्तारश्च प्रसादोऽथ व्यावृत्तस्खलितावपि ।
परिवर्ताक्षेपबिन्दूद्वाहितोर्मिसमास्तथा ॥२७॥

प्रेङ्खनिष्कूजितश्येनक्रमोद्धट्टितरञ्जिताः ।
संनिवृत्तप्रवृत्तोऽथ वेणुश्च ललितस्वरः ॥२८॥

हुङ्कारो ह्लादमानश्च ततः स्यादवलोकितः ।
स्युः सञ्चारिण्यलङ्काराः पञ्चविंशतिरित्यमी ॥२९॥

Sanchari Alankara		The examples used are Arohi here. As Sanchari are circular in nature, if these are reversed in Descending order then it becomes Avrohi Sanchari Alankara	English notes without higher lower symbols.
1	Mandradi	Here there are five phrase (kala) of three tones each. Lower (Mandra) tone is progressive and is the first tone in the three notes in each of the five phases. This progressive Mandra first tone is the base to form the phrase. The third note is placed in the middle and sequence is as follows; 1-3-2, 2-4-3, 3-5-4, 4-6-5, 5-7-6	s-g-r r-m-g g-p-m m-d-p p-n-d
2	Mandramadhya	Here there are five phrase (kala) of three tones each. Lower (Mandra) tone is progressive and is middle tone in the three notes in each of the five phases. This progressive tone is the base to form the phrase. The first note of natural order of phrase is placed in the middle and sequence is as follows; 3-1-2, 4-2-3, 5-3-4, 6-4-5,7-5-6	s-g-r r-m-g g-p-m m-d-p p-n-d
3	Mandranta	Here there are five phrase (kala) of three tones each. Lower (Mandra) tone is progressive and is Last tone in the three notes in each of the five phases. This progressive tone is the base to form the phrase. The first Mandra note of natural order of phrase is placed in the end of the three notes and sequence is as follows; 2-3-1, 3-4-2, 4-5-3, 5-6-4,6-7-5	r-g-s g-m-r m-p-g p-d-m d-n-p
4	Prastara	Extension of phrases where a pair of notes is formed by dropping tone in between them and next pair is initiated by the omitted note in earlier one. Series becomes; 1-3, 2-4, 3-5, 4-6, 5-7.	s-g r-m g-p m-d p-n
5	Prasada	Here the phrases are also three notes. Every phrase has three tones and the 'next' tone will be placed in middle sandwiched between the preceding tone. 'Next' tone is progressive. So it will be e.g. s-r-s. Series numerically is 1-2-1, 2-3-2, 3-4-3, 4-5-4, 5-6-5, 6-7-6	s-r-s r-g-r g-m-g m-p-m p-d-p d-n-d
6	Vyavritta	This is a 5 note phrase. Here the sequence of 5 tones is 1-3-2-4-1 and so on. So the first two tones proceed from 1 to 3 and the next two from 2 to 4 ending back again with 1. Next series drops the first note and progresses making this order; 1-3-2-4-1, 2-4-3-5-2, 3-5-4-6-3, 4-6-5-7-4	s-g-r-m-s r-m-g-p-r g-p-m-d-g m-d-p-n-m
7	Skhalita	Here we start with Mandradi phrase and add 'next' note at end THEN we descend also in same phrase back to initial note. So sequence is 1-3-2-4-4-2-3-1, 2-4-3-5-5-3-4-2, 3-5-4-6-6-4-5-3, 4-6-5-7-7-5-6-4. TIP: another way to think of it is that this is it is started like Vyavritta where the first two tones proceed from 1 to 3 and the next two from 2 to 4 and instead of ending with 1 in Vyvritta here there are only four main tones and then the same phrase reverses as in {(1-3-2-4)-(4-2-3-1)}	s-g-r-m-m-r-g-s r-m-g-p-p-g-m-r g-p-m-d-d-m-p-g m-d-p-n-n-p-d-m

8	Parivartak	Here the first phrase (kala) is of three tones and the second tone (of natural order) is skipped in the first phrase. Then next phrase starts with omitted tone and proceeds with same logic. The relationship is 1-3-4, 2-4-5, 3-5-6, 4-6-7	s-g-m r-m-p g-p-d m-d-n
9	Akshepa	Here three note phrase is used and the sequence proceeds in ascending order but each progressive sequence drops the first tone of the preceding sequence forming; 1-2-3, 2-3-4, 3-4-5, 4-5-6, 5-6-7	s-r-g r-g-m g-m-p m-p-d p-d-n
10	Bindu	Here tones are deployed in phrases of three notes; triple prolonged (pluta) first note followed by short (Hrsva) next note and ending again with short first tone forming 1(3)-2-1, 2(3)-4-2, etc... prolonged.	s3-r-s r3-g-r g3-m-g m3-p-m p3-d-p d3-n-d
11	Udvahita	This is a four note phrase where having sung the first three tones the second tone repeats at end. The first tone is dropped in subsequent phrases forming; 1-2-3-2, 2-3-4-3, 3-4-5-4, 4-5-6-5, 5-6-7-6	s-r-g-r r-g-m-g g-m-p-m m-p-d-p p-d-n-d
12	Urmi	This is a four note phrase with structure. First note of murchana-fourth note (pluta) - first note - fourth note. Thus 1-4(3)-1-4, 2-5(3)-2-5, 3-6(3)-3-6, 4-7(3)-4-7	s-m3-s-m r-p3-r-p g-d3-g-d m-n3-m-n
13	Sama	Here phrases are four tones each followed by reverse of the initial four tones i.e. 1-2-3-4-4-3-2-1, 2-3-4-5-5-4-3-2, 3-4-5-6-6-5-4-3, 4-5-6-7-7-6-5-4. Due to the symmetry of ascent and descent in same phrase it is called Sama Alankara	s-r-g-m-m-g-r-s r-g-m-p-p-m-g-r g-m-p-d-d-p-m-g m-p-d-n-n-d-p-m
14	Prenkha	Word 'Prenkha' in Sanskrit means a swinging cradle. So as the swinging motion back and forth here the four tones in the phrase move forward and backward. Each 4 tone phrase deploys two tones and proceeds as such; 1-2-2-1, 2-3-3-2, 3-4-4-3, 4-5-5-4, 5-6-6-5, 6-7-7-6	s-r-r-s r-g-g-r g-m-m-g m-p-p-m p-d-d-p d-n-n-d
15	Nishkujita	This is five notes long and is an extension of Prasada (which deals with two tones) which is three notes long. After the Prasada, one progresses to third tone here and then reverts back to initial tone as a the fifth note. Forming sequences; 1-2-1-3-1, 2-3-2-4-2, 3-4-3-5-3, 4-5-4-6-4, 5-6-5-7-5	s-r-s-g-s r-g-r-m-r g-m-g-p-g m-p-m-d-m p-d-p-n-p
16	Shyena	This is Alankara formed by using consonant Shadaj-Pancham bhava series; note pairs which are 9 shrutis apart. Notes used here are s, r, g and m as initial notes.	s-p r-d g-n m-s'

281

17	Krama	Here it is sequential flow of three phrases (kala), each phrase begins with initial note and is 2,3 and 4 notes long respectively forming; 1-2-1-2-3-1-2-3-4, 2-3-2-3-4-2-3-4-5, 3-4-3-4-5-3-4-5-6, 4-5-4-5-6-4-5-6-7	s-r-s-r-g-s-r-g-m r-g-r-g-m-r-g-m-p g-m-g-m-p-g-m-p-d m-p-m-p-d-m-p-d-n
18	Udghatita	Here phrases are six notes long. After singing first two tones, one jumps skips first to fifth tone counted from the initial note and then descends up to the second note. Thus forming; 1-2-5-4-3-2, 2-3-6-5-4-3, 3-4-7-6-5-4. Other way to think of is after singing the first two notes one descends from the consonant of the first note. s-p, r-d and g-n pairs as 1st and third tones.	s-r-p-m-g-r r-g-d-p-m-g g-m-n-d-p-m
19	Ranjita	This is based on Mandradi discussed in number 1 here. Sequence is (Mandradi)x2-initial Mandra lower note at end. Forming; 1-3-2-1-3-2-1, 2-4-3-2-4-3-2, 3-5-4-3-5-4-3, 4-6-5-4-6-5-4, 5-7-6-5-7-6-5	s-g-r-s-g-r-s r-m-g-r-m-g-r g-p-m-g-p-m-g m-d-p-m-d-p-m p-n-d-p-n-d-p
20	Sannivrtapravartak	Here one sings first two notes in ascending order and the first two notes are consonant pairs s-p. Then one descends three notes from the fourth to the second. One proceeds by dropping initial note and forms other phrases. Numerically; 1-5-4-3-2, 2-6-5-4-3, 3-7-6-5-4	s-p-m-g-r r-d-p-m-g g-n-d-p-m
22	Venu	Here first note is sung twice then order is followed of second, fourth and third forming a five note phrase. Proceed by dropping initial note and incrementing. Numerically; 1-1-2-4-3, 2-2-3-5-4, 3-3-4-6-5, 4-4-5-7-6	s-s-r-m-g r-r-g-p-m g-g-m-d-p m-m-p-n-d
23	Lalitsvara	This is a five tone phrase where first tow tones are sung followed by 4th and the descending the same first two notes. Numerically; 1-2-4-2-1, 2-3-5-3-2, 3-4-6-4-3, 4-5-7-5-4	s-r-m-r-s r-g-p-g-r g-m-d-m-g m-p-n-p-m
24	Hunkara	This Alankara grows sequentially in phrase length where first two notes are sung and then descend to initial note. Then we add the third note and descend sequentially forming his series; 1-2-1, 1-2-3-2-1, 1-2-3-4-3-2-1, 1-2-3-4-5-4-3-2-1, 1-2-3-4-5-6-5-4-3-2-1	s-r-s s-r-g-r-s s-r-g-m-g-r-s s-r-g-m-p-m-g-r-s s-r-g-m-p-d-p-m-g-r-s s-r-g-m-p-d-n-d-p-m-g-r-s
25	Hradamana	This is related to Mandradi and Ranjita. It is half in size of Ranjita where Mandradi id repeated twice and ends again with the Mandra lower initial tone. Here Mandradi is sung once followed immediately by the lower tone. Numerically; 1-3-2-1, 2-4-3-2, 3-5-4-3, 4-6-5-4, 5-7-6-5	s-g-r-s r-m-g-r g-p-m-g m-d-p-m p-n-d-p
26	Avalokita	This is related to Sama. Where the four tone Sama drops the second tone in both ascent and descent it becomes Avalokita. 1-3-4-4-2-1, 2-4-5-5-3-2, 3-5-6-6-4-3, 4-6-7-7-5-4.	s-g-m-m-r-s r-m-p-p-g-r g-p-d-d-m-g m-d-n-n-p-m

7 Extra Embellishments

(ऋ) सप्तान्यलंकारोद्देश·

अन्ये ऽपि सप्तालंकारा गीतज्ञैरुपदर्शिताः ।
तारमन्द्रप्रसन्नश्च मन्द्रतारप्रसन्नकः ॥५४॥

आवर्तकः सम्प्रदानो विधूतो ऽप्युपलोलकः ।
उल्लासितश्चेति तेषामधुना लक्षम कथ्यते ॥५५॥

	7 Extra Alankara	In each of these the first phrase is initiating phrase and all subsequent phrases are formed by dropping initial note of the previous phrase.	Hindi with Higher lower notes symbols denoted.	English notes without higher lower symbols.
1	Taramandra-Prasanna	Here one ascends to the eight note in the higher register (Tara saptak) and then ends back with initial note (Mandra). Thus forming; 1-2-3-4-5-6-7-8-1, 2-3-4-5-6-7-8-9-2 and so on...	s-r-g-m-p-d-n-s'-s r-g-m-p-d-n-s'-r'-r ...	s-r-g-m-p-d-n-s'-s r-g-m-p-d-n-s'-r'-r ...
2	Madratara-Prasanna	Here one sings first note Mandra and then jumps to eighth note in (Tara) and descends back to Mandra initial note. Thus forming; 1-8-7-6-5-4-3-2-1, 2-9-8-7-6-5-4-3-2....	s-s'-n-d-p-m-g-r-s r-r'-s-n-d-p-m-g-r ...	s-s'-n-d-p-m-g-r-s r-r'-s-n-d-p-m-g-r ...
3	Avartaka	Here first and second note are repeated twice followed by initial note again twice followed by second and ending in first. 1-1-2-2-1-1-2-1, 2-2-3-3-2-2-3-2, 3-3-4-4-3-3-4-3, 4-4-5-5-4-4-5-4, 5-5-6-6-5-5-6-5,	s-s-r-r-s-s-r-s r-r-g-g-r-r-g-r g-g-m-m-g-g-m-g m-m-p-p-m-m-p-m p-p-d-d-p-p-d-p d-d-n-n-d-d-n-d	s-s-r-r-s-s-r-s r-r-g-g-r-r-g-r g-g-m-m-g-g-m-g m-m-p-p-m-m-p-m p-p-d-d-p-p-d-p d-d-n-n-d-d-n-d
4	Sampradana	This is same as Avartaka minus the last two single tones.	s-s-r-r-s-s r-r-g-g-r-r g-g-m-m-g-g m-m-p-p-m-m p-p-d-d-p-p d-d-n-n-d-d	s-s-r-r-s-s r-r-g-g-r-r g-g-m-m-g-g m-m-p-p-m-m p-p-d-d-p-p d-d-n-n-d-d
5	Vidhuta	Here pairs of alternate notes are used twice in first phrase. Next phrase starts with the omitted tone in same manner. 1-3-1-3, 2-4-2-4, 3-5-3-5, 4-6-4-6, 5-7-5-7	s-g-s-g r-m-r-m g-p-g-p m-d-m-d p-n-p-n	s-g-s-g r-m-r-m g-p-g-p m-d-m-d p-n-p-n
6	Upalola	Here first two notes are repeated twice followed by third and second note pair twice. Looking like; [(1-2)x2-(3-2)x2] sequentially, 1-2-1-2-3-2-3-2, 2-3-2-3-4-3-4-3, 3-4-3-4-5-4-5-4, 4-5-4-5-6-5-6-5, 5-6-5-6-7-6-7-6	s-r-s-r-g-r-g-r r-g-r-g-m-g-m-g g-m-g-m-p-m-p-m m-p-m-p-d-p-d-p p-d-p-d-n-d-n-d	s-r-s-r-g-r-g-r r-g-r-g-m-g-m-g g-m-g-m-p-m-p-m m-p-m-p-d-p-d-p p-d-p-d-n-d-n-d
7	Ullasita	Here first note is repeated twice and followed by third, first and third again once. Repeat by omitting the first note. 1-1-3-1-3, 2-2-4-2-4, 3-3-5-3-5, 4-4-6-4-6, 5-5-7-5-7	s-s-g-s-g r-r-m-r-m g-g-p-g-p m-m-d-m-d p-p-n-p-n	s-s-g-s-g r-r-m-r-m g-g-p-g-p m-m-d-m-d p-p-n-p-n

Purpose and Object of Tonal Embellishments

(iii) अलंकाराणां प्रयोजनम्

रक्तिलाभः स्वरज्ञानं वर्णाङ्गानां विचित्रता ।।६४।।

इति प्रयोजनान्याहुरलङ्कारनिरूपणे ।

।। इति प्रथमे स्वरगताध्याये षष्ठं वर्णलङ्कारप्रकरणम् ।।६।।

The objective of the Varnalankara are threefold.

a) First, they assist the student in producing delightful music; Melody that touches one's soul.

b) Secondly, it teaches one the knowledge of the tonal perception 'Swara Gyan'. By singing the Alankara repeatedly, the tonality of melody gets seep into the being of the singer.

c) Finally, the Varnalankar as the name suggests makes one intimately family with the 'Varna' types and the tonal structural patterns.

The key to understand is that these Varnalankara are the practical pathways to understanding all the melodious science and process discussed so far. Every aspect of a melody introduced in earlier chapters can only be 'experienced' through practical application and that process starts with the Alankara. How should one practice them? Select the one that suits the practitioner and try and repeat the 'varna' to go into the inner essence of the melody, there will be modes and turns in the structures that your soul might find instinctively pleasing relish those and continue onto next phrases. Slowly with practice, each phrase 'Varna' will 'talk' to you it is when one feels that connectedness to the 'Varna' that one is said to have arrived at the point of

'Swar Gyan'- experiential practical knowledge of the tonality and the notes.

Jati

Jati Definition

We now introduce the 'applied' aspect of melody called 'Jati'. These are 'fundamental' melodic types on basis of which all other forms of practical (vocal and instrumental) melodic structures are developed.

The initial discussion starts with understanding what are Jatis and their thirteen main characteristics based on which eighteen types of Jatis are defined.

In simple terms Jatis are tonal structures used for the practical purpose of vocal singing and melodic reproduction on instruments. The earliest definitions of Jatis were defined systematically by Bharata in his Natya Shashtra treatise and were of eighteen types. These same eighteen Jatis form the bases of all further melodic structure creation practically and are referenced more or less in the same manner by other scholars since Bharata.

To further understand 'Jatis' one must understand its Sanskrit linguistic origin which signifies genus, class or a type of similar grouping. There are common attributes and characteristics of the group that all members possess which identifies them as part of this group, but the individual members also have their own unique characteristics that identify them from other individuals. This is akin to saying all Humans are a 'Jati' within a larger category of mammals and all humans are uniquely differentiated by their DNA yet they are humans.

'Jatis' in terms of music possess thirteen key characteristics that define the uniqueness and common elements of the eighteen overall types of Jatis. In music, we define these melodic tonal structures as Jatis because they pertain to a 'group' of collective factors such as Shruti, Swara, Grama and their starting and ending notes, etc. Thus, they form a group of tonalities that evoke a certain emotional mood "rasa" that uniquely defines a given Jati. *In other words, Jatis form the base foundation of the entire 'Raga' system.*

| | | | | Modifications * | | | Formed by | Hexatonic / Pentatonic Possible |
#	Name	Grama	Shudha Modified	Mod. For m	Loss of Complete	Violating rule	Combination of Shudh Jatis	Note Compl. =7 Shadav = 6 Audav = 5
1	Shadaji	Shadaj	Shudh	15	8	5	NA	7,6
2	Shadaja Kaishiki	Shadaj	Modified				Shadaji, Gandhari	7
3	Shadajotyava	Shadaj	Modified				Shadaji, Gandhari, Dhaivati	7,6,5
4	Shadaj Madhyam	Shadaj	Modified				Shadaji, Madhyama	7,6,5
5	Arshbhi	Shadaj	Shudh	23	16	7	NA	7,6,5
6	Dhaivati	Shadaj	Shudh	23	16	7	NA	7,6,5
7	Naishadi	Shadaj	Shudh	23	16	7	NA	7,6,5
8	Gandhari	Madhyam	Shudh	23	16	7	NA	7,6,5
9	Madhyama	Madhyam	Shudh	23	16	7	NA	7,6,5
10	Panchami	Madhyam	Shudh	23	16	7	NA	7,6,5
11	Gandharoditchyava	Madhyam	Modified				Gadhari, Dhavati, Shadaji & Madhyama	7,6
12	Raktgandhari	Madhyam	Modified				Gandhari, Naishadi, Panchami & Madhyama	7,6,5
13	Kaishiki	Madhyam	Modified				Shadaji, Gandhari, Madhyama, Panchami & Naishadi	7,6,5
14	Madhyomidtchyava	Madhyam	Modified				Gandhari, Dhaviati, Panchami & Madhyama	7
15	Karmaravi	Madhyam	Modified				Naishadi, Panchami & Arshbhi	7
16	Gandharpanchami	Madhyam	Modified				Gandhari, Panchami	7
17	Andhri	Madhyam	Modified				Gandhari, Arshbi	7,6
18	Nandayanti	Madhyam	Modified				Gandhari, Panchami & Arshbhi	7,6

* Violating rules of initial (graha), fundamental (ansha) or semifinal (apnyasa) Swara

18 Jatis

#	Name	Hexatonic / Pentatonic Possible Note Compl. =7 Shadav = 6 Audav = 5	Ommited Notes		Final	Fundamental Notes			Semi Final Notes	
			Shadav = 6 ommited note	Audav = 5 ommited note	Nyas Swar	# of Funda-mental Notes	(Ansha Swara) Fundamental Notes	# of Semi-Final Notes	(Apnyasa) Semi-Final Notes	
1	Shadaji	7,6	n	NA	s	5	s g m p d	2	g p	
2	Shadaja Kaishiki	7	NA	NA	g	3	s g p	3	s n p	
3	Shadajotyava	7,6,5	r	r p	m	4	s m d n	2	s d	
4	Shadaj Madhyam	7,6,5	n	n g	s m	7	s r g m p d n	7	s r g m p d n	
5	Arshbhi	7,6,5	s	s p	r	3	r n d	3	r n d	
6	Dhaivati	7,6,5	p	p s	g	2	r d	3	r m d	
7	Naishadi	7,6,5	p	p s	m	3	n r g	3	n r g	
8	Gandhari	7,6,5	r	r d	p	5	s g m p n	2	s p	
9	Madhyama	7,6,5	g	g n	d	5	s r m p d	5	s r m p d	
10	Panchami	7,6,5	g	g n	n	2	r p	3	n r p	
11	Gandharoditchyava	7,6	r	NA	A	2	s m	2	s d	
12	Raktgandhari	7,6,5	r	r d	g	5	s g m p n	1	m	
13	Kaishiki	7,6,5	r	r d	n p g	6	s g m p d n	6 or(7) s g m p d n (ri)	
14	Madhyomidtchyava	7	NA	NA	m	1	p	2	s d	
15	Karmaravi	7	NA	NA	p	4	r p d n	4	r p d n	
16	Gandharpanchami	7	NA	NA	g	1	p	2	r p	
17	Andhri	7,6	s	NA	g	4	r g p n	4	r g p n	
18	Nandayanti	7,6	s	NA	g	1	p	2	m p	

13 Lakshana of Jatis

ख. जातीनां त्रयोदश-सामान्यलक्षणानि

ग्रहांशतारमन्द्राश्च   न्यासापन्यासकौ   तथा ।
अपि संन्यासविन्यासौ बहुत्वं चाल्पता ततः ॥२९॥
एतान्यन्तरमार्गेण   सह   लक्ष्माणि   जातिषु ।
षाडवौडुविते   क्वापीत्येवमाहुस्त्रयोदश ॥३०॥

Thirteen Lakshana (characteristics) of Jatis are identified. The combination and permutation of these thirteen when grouped in a specific manner with other tonal aspects such as Shruti, Swara and Grama form an individual Jati melodic structure.

Prior to discussing the types of Jatis one must familiarize oneself with these thirteen characteristics which we will define one by one now. (29-30)

The earliest definitions of characteristics of Jatis were defined systematically by Bharata in his Natya Shashtra treatise and were TEN. However, some scholars in later periods like Sharangdev consider three elements separately thus totaling thirteen. As there are ten pranas of Taal and in this book, we have discussed Ten pranas of Melody it would be natural to define the Ten pranas of Jatis themselves which are their Lakshana. However being that we are relying on Sharangdev's treatise as our musical base, we will list the other three as well.

The original format considered Samnaysa and Vinyasa as part of Apnyasa characteristics since both of them define separate sections of a composition. Therefore, while we list them we have changed the order as a subcategory under Apnyasa to still identify the original Ten Prana of Jatis. Similarly, Antarmarga signifies a mutual relationship between a fundamental note and other factors and hence is shown under Fundamental note. The list is as follows;

1. Graha – Initial note

2. Ansha – Fundamental note

   a. Antarmarga

3. Tara – High pitch

4. Mandra – Low pitch

5. Nyasa – Final note

6. Apnyasa – Semifinal note

    a.   Samnyasa

    b.   Vinyasa

7.   Bahutva – Profusion

8.   Alpatva – Rareness

9.   Shadav – Hexatonic

10. Audav - Pentatonic

Graha - Initial Note

१· ग्रह

गीतादिविनिहितस्तत्र स्वरो ग्रह इतीरितः ।

तत्रांशग्रहयोरन्यतरोक्ताबुभयग्रहः ॥३१॥

1. The initial note (*graha*) : (31)

This is the very first initial note when commencing the melodic structure. In practice, the initial note is the same as the fundamental note of the Jati and hence the terms are often used interchangeably. In other words, the fundamental (Ansha) note that we define next is always the initial (Graha) swara of the melody. Thus, whenever either of these terms is indicated they mean the same through the process of implication. Graha literally means the place of residence from which all journeys start. The Graha swar is hence the initial first note from which all other aspects of melody evolve in the composition.

There are 63 fundamental (Ansha Swara) notes and hence there are said to be same 63 Graha Swara. (31)

Ansha - Fundamental note

2. अंशः

यो रक्तिव्यञ्जको गेये यत्संवाद्यनुवादिनौ ।
विदार्यां बहुलौ यस्मात्तारमन्द्रव्यवस्थितिः ॥३२॥

यः स्वयं यस्य संवादी चानुवादी स्वरो ऽपरः ।
न्यासापन्यासविन्याससंन्यासप्रहतां गतः ॥३३॥

प्रयोगे बहुलः स स्याद्धादशांशां योग्यतावशात् ।
बहुलत्वं प्रयोगेषु व्यापकं त्वंशलक्षणम् ॥३४॥

2. The fundamental note *(amsa)*: (32-34)

This is the most prominent fundamental note of the melodic structure. This note expresses the delightfulness of the vocal (includes instrumental) format of melodic structure. The various aspect of 'Ansha' swara and its definition are;

    i. It is the note that most significantly expresses the tonal emotion of the melodic structure in its delightfulness.

    ii. Vidari is the sub-section of melodic composition and the Ansha Swara is the note whose consonants (Vadi) and assonants (Samvadi) profusely expand the melodic dialogue in the vidari.

    iii. This is the note that determines higher (Tara) and lower (Mandra) sections of the composition

    iv. It is always its own consonant but can have another note as its assonant

    v. Ansha swara can become Graha swara, Nyasa, Apnyasa, Samnyasa, Vinyasa

    vi. It is most often used in the melodic structure

    vii. Due to its ability to delight and to be relished in melodic form, it is the note which is sonant (Vadi)

viii.   For all practical purposes, the note that is most profusely used (Bahulatvam) is the fundamental Ansha Swara.

(32-34)

As indicated earlier, in practice the initial note is the same as the fundamental note of the Jati and hence the terms are often used interchangeably. However, the theoretical difference is as follows;

1.   The difference between the two is functional in nature.

2.   Fundamental note acts as the 'Vadi' Sonant note of the composition.

3.   Graha swara is fourfold in nature.

4.   Ansha (fundamental) swara is of primary importance as it determines and acts as an originator of the Raga melodic structures.

5.   Graha swara is secondary in nature.

The 'Ansha' literally in Sanskrit means a constituent part of a larger entity. This note thus defines the constituent sub-sections (vidari) of the composition. Also, the Ansha swara is not only delightful on its own as are all other notes, but it is 'fundamental' because it expresses delightfulness of the complete relationship of all other notes with it in the composition.

Other special characteristics of Ansha swar are the fact that it can act as its own consonant 'Samvadi' note. This is seen in ancient pre-Murchana ragas such as Marva, Puriya and Gurjari and Sohini.

The Ansha swara also can act as semi-final (Vinyasa) or final note (Samnyasa) and thus it expands and profuse in the melody. Sage Bharata defines this note as "the note on which the charm of the Raga hinges upon and from which the charm profuses on basis of traversing the lower and higher pitches.

Antarmarga

11. अन्तरमार्गः

न्यासादिस्थानमुज्झित्वा मध्ये मध्ये ऽल्पतायुजाम् ।
स्वराणां या विचित्रत्वकारिण्यंशादिसंगतिः ।।५२।।
अनभ्यासैः क्वचित्क्वापि लङ्घनैरेव केवलैः ।
कृता सा ऽन्तरमार्गः स्यात्प्रायो विकृताजातिषु ।।५३।।

This is a special rare melodic artistic usage of a note that is used in its relationship to the 'Ansha' swara in modified Jatis. When one leaves the initial, fundamental, semifinal and final notes as it is including the Samnyasa and Vinyasa notes and introduces rare note in between in very limited duration in consonance with the 'Ansha' swara one creates new modified melodic structures. This is the process of 'Antarmarga' used in modified Jatis. (52-53)

Tara - Higher pitch range

३. तार:

मध्यमे सप्तके ऽंश: स्यात्तस्मात्तारस्थितात्परान् ।
स्वरांश्रतुर आरोहेदेष तारावधि: पर: ॥३५॥

अर्वाक्तु कामचार: स्यात्तारे लुप्तो ऽपि गण्यते ।
आतारषड्जमारोहो नन्दयन्त्यां प्रकीर्तित: ॥३६॥

Now we define the higher pitch limits of the melodic types. The seven notes of the middle register form the middle heptad. The fundamental note is always considered middle heptad and the upper range of the pitch is determined by the distance of four notes from the fundamental note in the upper heptad. This distance of four notes includes eliminated notes in the composition if any. E.g. in Shadaj Grama if fundamental notes is Shadaj, then in Tara Saptak (higher pitch heptad) one can go up to four notes from Tara Shadaj up to Tara Pancham.

However, in Madhyam Grama if Madhyam is the Ansha swar then one can proceed up to the Tara Nishad in upper heptad because the Tara Shadaj of the third (Mandra, Madhyam, and Tara are the three heptad pitch ranges) heptad is the final limit and is not considered. (35-36)

Mandra- Lower pitch

४. मन्द्र:

मध्यस्थानस्थितादंशादामन्द्रस्थांशमाव्रजेत् ।
आमन्द्रन्यासमथवा तदध:स्थरिधाबपि ॥३७॥

एषा मन्द्रगते: सीमा ततोऽर्वाक्कामचारिता ।

Now we define the lower (Mandra) pitch limits of the melodic types. The seven notes of the middle register form the middle heptad. The fundamental note is always considered in the middle heptad. From the fundamental note in middle heptad one can descend up to the fundamental note in Mandra saptak or up to the lower final 'Nyasa' swara and in some cases even below to Rishabh or Dhaivat of they are below. Thus, there are threefold possibilities of traversing the lower register profusely. (37)

Nyasa - Final note

5. न्यास:

गीते समाप्तिकृन्यास एकर्विंशतिधा च सः ॥३८॥

षाड्ज्यादीनां तु सप्तानां न्यासः स्यान्नामकृत्स्वरः ।

द्वौ नामकारिणौ षड्जमध्यमायां तु तौ मतौ ॥३९॥

उदीच्यवात्रयं मान्तं निपगान्ता तु कैशिकी ।

कार्मारवी पञ्चमान्ता गान्ताः पञ्चापराः स्मृताः ॥४०॥

This is the very LAST note when concluding the melodic structure. This happens in 21 varieties. For Shuddha Jatis, the Nyasa swar is their main denominative note. Of the 18 Jatis, the Shadaj-Madhyam Jati has two final note possibilities, Kaishiki has three and the remaining sixteen have one final note possibility. Refer to Jati Table. (39-40)

Apnyasa - Semifinal note

6. अपन्यासः

अपन्यासस्वरः स स्याद्यो विदारीसमापकः ।
कार्मारव्यां च नैषाद्यामान्ध्रीमध्यमयोस्तथा ॥४१॥

आर्षभ्यां च स्वरा ये ड्ंशास्ते ऽपन्यासाः प्रकीर्तिताः ।
उदीच्यवानां त्रितये ऽपन्यासो षड्जधैवतौ ॥४२॥

मध्यमो रक्तगान्धार्या गान्धार्या षड्जपञ्चमौ ।
सनिपाः षड्जकंशिक्यां पञ्चम्यां निरिपाः स्मृताः ॥४३॥

रिपौ गान्धारपञ्चम्यां बाड्ज्यां गान्धारपञ्चमौ ।
धैवत्यां रिमधाः प्रोक्ता नन्दयन्त्यां मपौ मतौ ॥४४॥

रिवज्र्याः षट् च कैशिक्यां सप्तापीत्यूचिरे परे ।
सप्तस्वरापन्यासां तु भावन्ते षड्जमध्यमाम् ॥४५॥

अत्र येऽ्ंशा अपन्यासास्ते स्युरेकोनविंशतिः ।
सप्तात्रिंशत्परे ते च षट्पञ्चाशत्तु संयुताः ॥४६॥

कैशिक्यां सप्तपक्षे तान्सप्तपञ्चाशतं विदुः ।

This is the note that concludes the sub-section (vidari) of composition and hence is called Apnyasa Swara (Semifinal note). Note that the composition has several constituent pieces and they are signified by terms signified by 'Dhatu' (Section), 'Vidari' (Sub-section) and 'Ang' (Part) of composition. Here we refer to the final note of the sub-section. Semifinal notes identical to fundamental notes are 19. The rest are 37 and totally they make 56. In Kaishiki, we count seven per some systems and hence the Apnyasa can be 57 in that case. Refer to the Jati table to see the mapping of Apnyasa swara to Jatis. (41-47)

Samnyasa

7. संन्यासः

अंशाविवादी गीतस्याद्यविदारीसमाप्तिकृत् ॥४७॥
संन्यासो ङ'शाविवादेव

'Samnyasa' is another specialized type of semi-final
note whereby it concludes the FIRST sub-section
(Vidari) of the composition. However, this note cannot
be dissonant (Vivadi) of the Ansha swar. (47)

Vinayasa

8. विन्यासः

, विन्यासः स तु कथ्यते ।
यो विदारीभागरूपपदप्रान्ते ङवतिष्ठते ॥४८॥

'Vinyasa' is another specialized type of semi-final note
whereby it concludes the end of a 'Pada' in the sub-
section of composition (Vidari). 'Pada' means a
completed word phrase, line of a stanza. A part or
portion of the overall poetic sentence. Just like
Samnyasa, this note cannot be dissonant (Vivadi) of the
Ansha swar. In other words, if the Samnyasa note
occurs at the end of the closing portion of the 'Pada' and
is NOT dissonant of the Ansha swar then it is called
Vinyasa. (48)

Bahutvam - Profusion

9. बहुत्वम्

अलङ्घनात्तथा ऽभ्यासाद्बहुत्वं द्विविधं मतम् ।
पर्यायांशे स्थितं तच्च वादिसंवादिनोरपि ॥४९॥

Next, we define two opposing but related characteristics of Jatis called Bahutvam (profusion) and Alpatvam (rareness). In case of Bahutvam, there are two varieties. One is achieved by NON-use of 'Langhanam'. Langhanam, which is related to the rareness concept, is translated as overstepping a note by just slightly touching it but still using the note. Therefore, in case of Bahutvam we do opposite of this and we do FULL touching and exploring the note and completely using the note as opposed to almost avoiding the note (in Alpatvam)

Second manner of Bahutvam is through use of full repetitions characterized by uninterrupted and frequent usage. This defined in Sanskrit by the very important term 'Abhyasa' which means mastering a concept by frequent and uninterrupted repetitions.

These two formats thus define Bahutvam. The one where the note is profuse by non-overstepping and fully touching the note is synonymous with Sonant 'Vadi' note. In addition, the Bahutvam through 'Abhyasa' repetitions leads to the consonant (Samvadi) swar. (49)

Alptava - Rareness

10. अल्पत्त्वम्

अल्पत्वं च द्विधा प्रोक्तमनभ्यासाच्च लङ्घनात् ।
अनभ्यासस्त्वनंशेषु प्रायो लोप्येष्वपीष्यते ॥५०॥
ईषत्स्पर्शो लङ्घने स्यात्प्रायस्तल्लोप्यगोचरम् ।
उशन्ति तदनंशे ऽपि क्वचिद्गीतविशारदाः ॥५१॥

Now in the case of Alpatvam (rareness) also there are two varieties. One is achieved by the absence of 'Abhyasa' (repetitions) of note and secondly by 'Langhanam', which we described above as overstepping a note by just slightly touching it but still using the note.

The 'Abhyasa' Alpatvam is never used for the Ansha swar of course but is used rather for the sonant (vadi) and co-fundamental notes. This technique also applies to omitted notes in cases of hexatonic and pentatonic types of Jatis.

'Langhanam' overstepping is used in case of omitted notes and sometimes for non-fundamental (any swara besides the Ansha) note. (50-51)

Shadav - Hexatonic

12. पाडवम्

षडवन्ति प्रयोगं ये स्वरास्ते पाडवा मताः ।
षट्स्वरं तेषु जातत्वाद्गीतं पाडवमुच्यते ॥५४॥

A rendering of melodic composition that is constituted of six swara omitting one note of the heptad is called hexatonic (Shadav). (54)

Audav - Pentatonic

13. औडुवम्

वान्ति यान्त्युडवो ऽत्रेति व्योमोक्तमुडुवं बुधैः ।
पञ्चमं तच्च भूतेषु पञ्चसंख्या तदुद्भवा ॥५५॥

औडुवी सा ऽस्ति येषां च स्वरास्ते त्वौडुवा मताः ।
ते संजाता यत्र गीते तदौडुवितमुच्यते ॥५६॥

तत्सम्बन्धादौडुवं च पञ्चस्वरमिदं विदुः ।
क्रमादल्पाल्पतरते षाडवौडुवकारिणो ॥५७॥

सम्पूर्णत्वदशायां स्तः, पञ्चम्यां तु बिपर्ययः ।
वचनं विधिरप्राप्ताविहाल्पत्वबहुत्वयोः ॥५८॥

परिसंख्या द्वयोः प्राप्तावेकस्यातिशयाय यत् ।

A rendering of melodic composition constituted from five swara omitting two notes of the heptad is called pentatonic (Audav). Sharangdev in case of Audav gives the root of the word as 'Udu' meaning the stars that dwell in the 'fifth' great element of Sky (ether) which is known in Sanskrit also as 'Uduvam'. This is the root of the name Audav meaning the composition leveraging five notes. (55-59)

Shuddha and Vikrit Jatis

Having understood the thirteen key characteristics of the Jatis, we now understand their two main categories of Pure and Modified format. These categories further define the 18 types of Jatis.

(i) शुद्धा जातयः

(क) सप्तशुद्धजातीनामुद्देशः

शुद्धाः स्युर्जातयः सप्त ताः षड्जादिस्वराभिधाः ।
षाड्ज्यार्षभी च गान्धारी मध्यमा पञ्चमी तथा ।।१।।
धंवती चाथ नैषादी,

(ख) शुद्धतालक्षणम्

, शुद्धतालक्षम कथ्यते ।
यासां नामस्वरो न्यासो ऽपन्यासो ंऽशो ग्रहस्तथा ।।२।।
तारन्यासविहीनास्ताः पूर्णाः शुद्धाभिधा मताः ।

(ii) विकृता जातयः

विकृता न्यासवर्जैतल्लक्षमहीना भवन्त्यमूः ।।३।।
सम्पूर्णत्वग्रहांशापन्यासेष्वेवंंवांऽवर्जनात् ।
भवन्ति भेदाश्चत्वारो द्वयोस्त्यागे तु षण्मताः ।।४।।
त्यागे त्रयाणां चत्वार एकस्त्यक्तं चतुष्टये ।
भेदाः पञ्चदशैवेते षाड्ज्याः सन्दूर्निरूपिताः ।।५।।

There are seven pure Jatis based on the seven notes. Pure (Shuddha) Jatis are defined as the ones where there Namaswara (denominative) note is the final (Nyasa) swara, semifinal (Apnyasa), fundamental (Ansha) and the initial (Graha) Swar. These cannot have a final note in Taar Saptak (higher register) and have complete seven notes. (1-2)

301

The above seven Pure (Shuddha) Jatis are defined as the ones where there Namaswara (denominative) note is the final (Nyasa) swara, semifinal (Apnyasa), fundamental (Ansha) and the initial (Graha) Swar. These cannot have final note in Taar Saptak (higher register) and have complete seven notes. When these qualities of Pureness are eliminated one by one from the Jati, then they become 'Modified' (Vikrit) in nature. The final note (Nyasa) swara still must be the Namaswara in Vikrit Jati.

By omitting completeness (seven notes), the initial note, fundamental note and semi-final, note one by one we get four varieties of modified Jatis. If we drop two of the above four criteria simultaneously we get six more modified Jatis. By omitting three aspects, we get four varieties and by omitting all four of the above aspects, we get one type of modified Jati. These form 15 types of Shadji Vikrit Jati.

Of the above 15, 8 are devoid of completeness and 7 are due to other factors. Of the ones that lack completeness, they are of two types Shadav (hexatonic) and Audav (pentatonic). Refer to the Jati Table above for reference of this categorization of Pure and Modified Jatis. (3-11)

Samsarja-Vikrata (Assoc. Modified Jatis)

(iii) संसर्गजा विकृता जातयः

विकृतानां तु संसर्गज्जाता एकादश स्मृताः ।
स्यात्स्वड्जकंशिको षड्जोदीच्यवा षड्जमध्यमा ॥८॥

गान्धारोदीच्यवा रक्तगान्धारी कंशिकी तथा ।
मध्यमोदीच्यवा कार्मारवी गान्धारपञ्चमी ॥९॥

तथा ऽऽन्ध्री नन्दयन्तीति,

तद्धेतूनधुना ब्रुवे ।
षाड्जीगान्धारिकायोगाज्जायते षड्जकंशिकी ॥१०॥

षाड्जिकामध्यमाभ्यां तु जायते षड्जमध्यमा ।
गान्धारीपञ्चमीभ्यां तु जाता गान्धारपञ्चमी ॥११॥

गान्धार्यार्षभिकाभ्यां तु जातिरान्ध्री प्रजायत ।
षाड्जी गान्धारिका तद्धद्धेवती मिलितास्त्विमाः ॥१२॥

षड्जोदीच्यवतीं जातिं कुर्युं:, कार्मारवीं पुनः ।
उत्पादयन्ति नैषादीपञ्चम्यार्षभिका युताः ॥१३॥

नन्दयन्तीं तु गान्धारीपञ्चम्यार्षभिका युताः ।
गान्धारी धंवती षाड्जी मध्यमेति युतास्त्विमाः ॥१४॥

गान्धारोदीच्यवां कुर्युर्मध्यमोदीच्यवां पुनः ।
एता एव विना षाड्जया पञ्चम्या सह कुबते ॥१५॥

कुर्युस्ता रक्तगान्धारीं नैषादो च न धंवती ।
अ.र्षभां धंवतीं त्यक्त्वा पञ्चम्यः कैशिकी भवेत् ॥१६॥

Jati Gram Classification

(iv) जातीनां ग्रामविभागः

चतत्रः षड्जशब्दिर्यो नैषादी धैवती तथा ।
आर्षभो चेति सप्तेताः षड्जग्रामस्य जातयः ॥१७॥
शेषाः स्युर्मध्यमग्रामे,

Four Jatis whose names start with Shadaj and Naishadi, Dhaivati and Arshbhi belong to the Shadaj Grama and remaining eleven (out of the total eighteen), Jatis belong to Madhyam Grama. Refer to the Jati Table above for reference to this categorization of Jatis grouped by Grama. (17)

Complete, Hexatonic and Pentatonic Jatis

(v) सम्पूर्ण-षाडवौडुव-जातयः

, पूर्णत्वाद्यधुनोच्यते ।
कार्मारव्यथ गान्धारपञ्चमो षड्जकंशिकी ॥१८॥
मध्यमोदीच्यवेत्येता नित्यपूर्णाः प्रकीर्तिताः ।
षाड्जी च नन्दयन्त्यान्ध्रो गान्धारोदीच्यवेत्यमूः ॥१९॥
सम्पूर्णंषाडवाः प्राह चतत्रः काश्यपो मुनिः ।
नशाब्ज्ञिष्ठाः सम्पूर्णंषाडबौडबिता मताः ॥२०॥

Depending on the number of notes in Jatis they can be classified as Hexatonic (Shadav – 6 notes), Pentatonic (Audav – 5 notes) or Complete (Seven notes). Refer to the Jati Table above for reference of this categorization of Jatis grouped by Note completion.

(vi) जातिषु स्वरसाधारणनियमः

पञ्चमीमध्यमाषड्जमध्यमाऽऽख्यासु   जातिषु ।
स्वरसाधारणं   प्रोक्तं   मुनिभिर्भरतादिभिः ॥२१॥

अंशेषु   समपेष्वेतद्यथास्वनियमाद्भवेत् ।
एतदल्पनिगास्वाहुः   कम्बलाश्वतरादयः ॥२२॥

अल्पद्विश्रुतिके   रागभाषाऽऽदावपि   तन्मतम् ।
निगयोरंशयोः   षड्जमध्यमायां न   तद्भवेत् ॥२३॥

विकृता   एव   तत्रापि   स्वरसाधारणाश्रयाः ।

As we discussed earlier in the section on Sadharana, it is a process of overlapping notes. This same Sadharan process, when applied to three Jatis, creates the Jati Sadharana or Overlapping of Notes in Jatis. The rules of which are as follows;

The overlapping can only occur in three Jatis called Panchami, Madhyama and Shadaj Madhyama. Here overlapping must be applied in relation to s, m and p notes which are to be used as fundamental notes in relation to the Jatis that have weak (alpa) g or n swar. Only modified (Vikrit) Jatis can employ Jati Sadharana. (21-23)

Number of Fundamental notes in Jatis

(vii) जातिगतांश्वरगणना

एकांशा नन्दयन्ती च मध्यमांदीच्यवा तथा ॥२४॥
गान्धारपञ्चमीत्येतास्तिस्रो द्वयंशास्तु पंबती ।
गान्धारोदीच्यवा चाथ पञ्चमीत्युदिता इमाः ॥२५॥
नैषाद्यार्षभिकाषड्जकैशिक्यस्त्र्यंशिका मताः ।
आग्धोकार्मारवीषड्जो दीच्यवाश्चतुरंशिकाः ॥२६॥
पञ्चांशा रक्तगान्धारी गान्धारी मध्यमा तथा ।
षाड्जीत्येताश्चतस्रः स्युः गडंशकैब कैशिकी ॥२७॥
सप्तांशा सूरिभिः षड्जमध्यमा परिकीर्तिता ।
इति त्रिषष्टिरंशाः स्युर्जातिष्वष्टादशस्विमे ॥२८॥

The total number of fundamental (Ansha) swara in the eighteen Jatis is 63 when one considers only complete Jatis. When Hexatonic (Shadav) Jatis are considered the total, possible fundamental notes can be 47. Similarly, it is observed that in Pentatonic (Audav) Jatis the number reduces to 30. Refer to the Jati Table above for reference to this categorization of Jatis grouped by the number of Fundamental notes allowed in it. (24-28)

Jati Concluding Instructions

घ. उपसंहारः

(i) जातिविषयकाः सामान्यनिर्देशाः

अनुक्ताविह तालः स्यात्रिधर्वैककलाऽविकः ।
मार्गाः क्रमाच्चित्रवृत्तिदक्षिणा, गीतयः पुनः ॥११०॥

मागधी संभाविता च पृथुलेत्युदिताः क्रमात् ।
योक्ता ऽस्माभिः कलासंख्या ता दक्षिणपथे स्थिता ॥१११॥

वार्तिके द्विगुणा ज्ञेया सैव चित्रे चतुर्गुणा ।
सर्वंजातिषु जानीयादंशस्वरगतं रसम् ॥११२॥

Now some basic fundamental assumptions of Jatis are summarized. When no Tala is indicated with a Jati, the following general rules about Tala, Marg, and Giti must be observed.

All Prastar of Jatis occurs in Dakshin Marg. The Kalas in the tala pertains to the Dakshin Marg. (Vartik Marg would have double and Chitra Marg would have quadruple Kala of Dakshin Marg). Time taken by the composition would be the same but the contents of the Kala would vary based on marg.

Jatis are to be set in Tala even when not specified and this tala must be threefold – Ekkala, Dvikala, and Chatushkala in the Chachput tala. For further details on the Tala and Kala system, one should refer to the 'Science of Rhythm' volume. The Gitis of Jatis when not specified are assumed Magadhi, Sambhavita and Prithula respectively

Aesthetic delight (Rasa) mood of Jatis is set by the fundamental (Ansha) swar in all Jatis. (110-112)

Sense, Purpose, and Objective of Jatis

(ii) जातिगानस्य फलश्रुतिः

ब्रह्मप्रोक्तपदैः   सम्यक्प्रयुक्ताः   शङ्करस्तुतो ॥११३॥
अपि   ब्रह्महणं   पापाज्जातयः   प्रपुनन्त्यमूः ।

ऋचो यजूंषि सामानि क्रियन्ते नान्यथा यथा ॥११४॥
तथा   सामसमुद्भूता   जातयो   वेदसंमिताः ।

Jatis originated from the Vedas and specifically they relate to the tonal structure and singing rules of Sama Gaan described in Samaveda. As such singing Jatis must be done with the same respect and care, as one must observe when singing the Vedas. Since Jatis are derived from Vedas, they are an 'Ansha' of the Vedas. The hymns of Rigveda, Yajurveda, Atharvaveda and the Samaveda (Four main Vedas) must be sung in their respective tonal rules and for Jatis to produce their true delight and other merits in both singer and listener, the rules must be understood and adhered to.

The physical benefit of mastering the Jatis is that they are the originators of all the Grama Ragas in the Indian Melodic system and hence the practitioner through inheritance, masters the Ragas automatically when one masters Jatis.

The esoteric benefit of Jatis is that due to their Vedic origins when one sings spiritual and devotional hymns composed in Jatis, the spiritual fruits are sure to be attained in a clearer manner. It is emphasized in the

Vedas that singing the hymns in reverence of the creator wash away the past sins of many lives and Jati Gaan is specifically suited for this objective. Sharangdev being a devotee of Lord Shiva gives an example of singing the hymns of Shiva as composed by Lord Brahma in the verse above. Moreover, some fruits are to be obtained by singing the glories focusing on one's own object of devotion.

Giti

From Jatis all other melodic types originate and
now we discuss main melodic types of
compositions (i.e. Giti or songs) that are derived
from Jatis in this final tenth prana of Melody.
These are the Gitis and they are of two types Kapala
Gana Giti and Kambala Gana Giti.

Gitis Definition and classification

(ii) गीतिलक्षणां, तन्द्रेदाश्च

वर्णद्यलङ्कृता गानक्रिया पदलयान्विता ॥१४॥

गीतिरित्युच्यते सा च बुधेरुक्ता चतुर्विधा ।

मागधी प्रथमा ज्ञेया द्वितीया चार्धमागधी ॥१५॥

संभाविता च पृथुलेत्येतासां लक्षम चक्ष्महे ।

(iii) पञ्च गीतयः

गीतयः पञ्च शुद्धा च भिन्ना गौडी च वेसरा ॥२॥ सी.

साधारणीति, शुद्धा स्यादबक्रैलंसितः स्वरैः ।

भिन्ना वक्रैः स्वरैः सूक्ष्ममंधुरंगंमफंर्युता ॥३॥

गाढंत्रिस्थानगमकरौहाटीलसितः स्वरैः ।

अखण्डितस्थितिः स्यानन्त्रये गौडी मता सताम् ॥४॥

ओहाटी फम्प्यिसंमंग्रेमुं बुबुसतरः स्वरैः ।

तुकारोकारयोगेन हृम्यस्ते बिबुके भवेत् ॥५॥

वेगवद्भिरू स्वरंबंणंचतुष्कोऽप्यतिरधिसतः ।

वेगस्वरा रागगीतिषंसरा खोच्यते शुर्षः ॥६॥

चतुर्गतिगसं लक्षम भिता साधारणी मता ।

शुद्धाविगीतियोगेन रागाः शुद्धावयो मता ॥७॥

The Giti are defined by the following characteristics;

1. Act of singing (Gana Kriya)

2. Embellished by the tonal patterns of Varnalankara

3. Composed of verbal phrases (Pada-poetic sentences)

4. Set in musical rhythmic tempo - Laya (drut, Madhya or Vilambit i.e. fast, medium or slow respectively)

There are three main types of Gitis.

1. Padashrita Giti – Act of singing the poetic phrases and words of Pada. These are of four subtypes.

   a. Magadhi (originated in ancient East Indian province of Magadha, hence the name)

   b. ArdhaMagadhi

   c. Sambhavita

   d. Prithula

2. Talashrita Giti – Act of singing the 'varna' of the rhythmic intonations (also called as parans) or alternatively act of singing the Padas bound by the specific Taal Matra combinations suggested by the Talashrita Giti.

3. Swarshrita Giti – Act of singing formations based on tonal peculiarities, which leads to the creation of the structured Ragas in their swar

notations. Sharangdev considers these as five types and some scholars consider them of seven types.

i.   Shuddha – includes captivating and straight (Avakra) swara.

ii. Bhinna – includes curved (Vakra or out of place) and fine swara.

iii. Gaudi – is characterized by swara that span and prevail across all three registers. Here a special type of singing style of swara in the lower register is employed known as 'Ohati' or 'AuHati'. Ohati is a technique where a low soft sound is produced in quick succession by leveraging a 'Gamak' known as Kampita deploying sounds of words 'Au' and 'Ha'. A Gamak is a shaken sound produced which is pleasing to the listener.

iv. Vesara (VegaSwara) – as name illustrates it deploys fast (Vega) notes across all four Varnas (tonal structures namely Sthayi, Sanchari, Arohi, and Avrohi) and is pleasing to the ear.

v.   Sadharani – combines features of all the four Swarshrita Gitis above.

vi. The last two Gitis also accepted by a scholar Matang are known as Bhasha and Vibhasha and they represent derived varieties of Grama based Ragas.

We briefly now describe these types of Gitis, as they are the final set of structured melodic foundation, which then leads us to the creative freedom of creating new Ragas. We will define Raga as a term in the summary of this volume.

Padashrita Giti

Note that Sharangdev in his discussion of 'Swara Shashtra' ends the chapter at Padashrita Giti and then switches to practical aspects of singing by introducing the concept of Ragas in separate section called 'Ragavivekadhyay' that are developed using the 'Swara Shashtra' described earlier. The Swarashrita Gitis are thus not discussed by Sharangdev with a chapter on Gitis but rather discussed in the section on Ragas in Sangeet Ratnakar and this is not in our present scope. Since these, Swarashrita Gitis also must be discussed in order to complete the analysis of Gitis, present Author includes the description of Swarshrita Gitis in this volume with all other Gitis. A detailed description of Padashrita Giti is given here as intended in the original version by Sharangdev. However, for those desirous to learn more about Ragas the entire volume on Sharangdev's 'Ragavivekadhyay' is recommended besides the numerous other Raga structural notation practice and theory guidebooks mentioned in the preface and in the bibliography.

Magadhi

1. मागधी गीतिः

गीत्वा कलायामाद्यायां विलम्बितलयं पदम् ॥१६॥

द्वितीयायां मध्यलयं तत्पदान्तरसंयुतम् ।
सतृतीयपदे ते च तृतीयस्यां द्रुते लये ॥१७॥

इति त्रिरावृत्तपदां मागधीं जगदुर्बुधाः ।

This Giti originated in Magadh province in ancient India and is named after this region. It is also known as Triavrutpada. If we recall the principle of creation with the triads in musical time and space dimension, this Giti is a practical example of deploying this triad concept.

Here three Laya of rhythmic tempo are used. Three Kala are used, the poetic sub-section phrase of the Pada is sung three times in the different Laya starting from Vilambit, Madhyam and Drut Laya in the third line of the Giti.

यथा—

मा	गा	मा	धा
बे		बं	
धनि	धनि	सनि	धा
बे	बं	ह	द्रं
रिग	रिग	मग	रिस
देवं	ह्द्रं	बं	दे

Thus, the structure of Magadhi Giti is as follows;

314

Padshrita Giti Triavrutapadam Magadhi Giti	Kala					Laya Ratio 1:4	Tempo
Swara	1	m	g	m	d	4	Vilambit	Slow
Pada	1	de	~	vam	~	4	Vilambit	Slow
Swara	2	dn	dn	sn	d	2	Madhyam	Medium
Pada	2	de	vam	ru	dram	2	Madhyam	Medium
Swara	3	rg	rg	mg	rs	1	Drut	Fast
Pada	3	devam	rudram	van	de	1	Drut	Fast

Ardha Magadhi

2. अर्धमागधी गीति:

पूर्बंयो: पदयोरर्धं चरमे द्विर्यंदोदिते ।।१८।।
तदा ड्र्धंमागर्धी प्राहु:

यथा—

मा	री	गा	सा
दे		बं	
सा	सा	धा	नी
वं	ऋ	द्रं	
पा	धा	पा	मा
द्रं	वं	दे	

द्विरावृत्तपदां परे ।

यथा—

मा	मा	मा	मा
दे		वं	
धा	सा	धा	नी
दे	वं ,	ऋ	द्रं
पा	निध	मा	मा
ऋ	द्रं	वं	दे

This is a modification of Magadhi Giti and first Kala of its structure is the same as Magadhi. Hence the name Ardha Magadhi. First Kala is the same as Magadhi. Then the second half of the first (Kalas) phrases are repeated at the beginning of the second Kala. Finally, the second part of the second Kala is repeated as first part of the Third Kala. The double repetition gives it another name as Dviavrutapada. This is clear by comparing examples to Magadhi above.

Padshrita Giti Triavrutapadam								
Magadhi Giti	**Kala**						**Laya Ratio 1:4**	**Tempo**
Swara	1	m	g	m	d	4	Vilambit	Slow
Pada	1	de	~	vam	~	4	Vilambit	Slow
Swara	2	dn	dn	sn	d	2	Madhyam	Medium
Pada	2	de	vam	ru	dram	2	Madhyam	Medium
Swara	3	rg	rg	mg	rs	1	Drut	Fast
Pada	3	devam	rudram	van	de	1	Drut	Fast
Dviavrutapadam							**Laya Ratio 1:4**	**Tempo**
ArdhMagadhi Giti Kala						4	Vilambit	Slow
Swara	1	m	r	g	s	4	Vilambit	Slow
Pada	1	de	~	vam	~	2	Madhyam	Medium
Swara	2	s	s	d	n	2	Madhyam	Medium
Pada	2	vam	ru	dram	~	1	Drut	Fast
Swara	3	p	dn	p	m	1	Drut	Fast
Pada	3	dram	van	de	~			
ArdhMagadhi Giti Kala							**Laya Ratio 1:4**	**Tempo**
Swara	1	m	g	m	d	4	Vilambit	Slow
Pada	1	de	~	vam	~	4	Vilambit	Slow
Swara	2	dn	dn	sn	d	2	Madhyam	Medium
Pada	2	de	vam	ru	dram	2	Madhyam	Medium
Swara	3	rg	rg	mg	rs	1	Drut	Fast
Pada	3	ru	dram	van	de	1	Drut	Fast

Sambhavita

3. सम्भाविता गीतिः

संक्षेपितपबा भूरिगुहः संभाविता मता ॥१९॥

यथा—

धा	मा	मा	रि	ग
भ		बत्या		
री	गा	सा	सा	
दे		बं		
नी	धा	सा	नी	
ष्ठ		द्रं		
धा	नी	मा	मा	
बं		दे		

This Giti is defined by abridged short phrases. Here the Padas are minimal but long sounded syllables are used. This is sung in Dvikala Chachput taal of Vartik Marg.

Sambhavita Giti	Kala				
Swara	1	d	m	m	rg
Pada	1	bha	~	ktya	~
Swara	2	r	g	s	s
Pada	2	de	~	vam	~
Swara	3	n	d	a	n
Pada	3	ru	~	dram	~
Swara	4	d	n	m	m
Pada	4	van	~	de	~

Prithula

4. पृथुला गीतिः

भूरिलध्वक्षरपदा पृथुला संमता सताम् ।

यथा—

मा	गा	री	गा
सु	र	न	त
सा	धनि	धा	धा
ह	र	प	ब
धा	सा	धा	नी
यु	ग	लं	
पा	निधप	मा	मा
प्र	ण	म	त

Many Padas with short syllables define this Giti. Therefore, the Padas are many here but syllable sounds are short here. This is sung in Chatushkala Chachput Taal of Dakshin Marg.

318

Prithula Giti	Kala				
Swara	1	m	g	r	g
Pada	1	su	ra	na	ta
Swara	2	s	dn	d	d
Pada	2	ha	ra	pa	da
Swara	3	d	s	d	n
Pada	3	yu	ga	lam	~
Swara	4	p	ndp	m	m
Pada	4	pra	na	ma	ta

Kapala and Kambala Giti

From the seven Shudha Jatis originate their corresponding Kapala and the Ragas that originate from these Jatis bear resemblance to their Kapala. These seven Kapala songs are said to have been composed by Lord Brahma in honor of Lord Shiva and practice and performance of these generate material prosperity and spiritual gain for the singer according to the author Sharangdev.

 We briefly describe here in a table, the seven Kapala Giti and one Kambala Giti and their characteristics. Note Kambala Giti is only one type and originates from Panchami Jati. Kambala in Vedic times were known to be the Naga (Serpent people), they created this Kambala form of devotional singing in honor of Lord Shiva, and since then it is known as Kambala Giti.

Kapala Giti Name	Kala	Initial Graha	Funda. Ansha	Final Nyasa	SemiFinal ApNyasa	Profuse Bahutvam	Rare Alptavam	Overstep Langhanam	Pentatonic Audav
Shadaji - Kapala	12	s	s	g	s	gm	rpnd gnpd	r	
Arshbhi - Kapala	8		r	m	r		s very rare		
Gandhari - Kapala	8	m	m	m	m	d	srg		rp
Madhyama - Kapala	9		m				nrgp		
Panchami - Kapala	8	s	r				nsgp		
Dhaivati - Kapala	8	s	d	p	s	md	rg	r	
Naishadi - Kapala	8	s	s	s		ndm	rg		

Kambala Giti

Name	Kala	Initial Graha	Funda. Ansha	Final Nyasa	SemiFinal ApNyasa	Profuse Bahutvam	Rare Alptavam	Overstep Langhanam	Pentatonic Audav
Panchami-Kambala		p	p	s	p	r	mdg		

2. आर्षभी-कपालम्

यत्रर्षभो ङ्शो ङपन्यासो मो ङन्तो गनिपधाल्पता ॥३॥
सो ङत्यल्पो ङष्टकलं तत्स्यात्कपालं त्वार्षभीगतम् ।

3. गान्धारी-कपालम्

मध्यमो ङ्शो ग्रहो न्यासो ङपन्यासो धैवतो बहुः ॥४॥
यत्राल्पाः सरिगा लोपाद्रिपयोरीडुबं भवेत् ।
तद्गान्धारीकपालं स्यात्कलाङष्टकविनिर्मितम् ॥५॥

4. मध्यमा-कपालम्

मध्यमो ङ्शो निरिगपाः स्वल्पा यत्र कला नव ।
तन्मध्यमाकपालं स्यादिति निःशङ्कसम्मतम् ॥६॥

5. पञ्चमो-कपालम्

ऋषभांशं सग्रहं च निधषड्जगमाल्पकम् ।
कपालं पञ्चमीजातिजातमष्टकलं विदुः ॥७॥

6. धैवती-कपालम्

अत्यल्पर्षभगान्धारं पन्यासं मधभूरि च ।
षाड्ज्या इव कपालं तद्धैवत्याः सकलाङष्टकम् ॥८॥

7. नैषादी-कपालम्

ग्रहांशन्यासषड्जं च रिगाल्पमतिभूरिभिः ।
निधमैरष्टकलकं स्यान्नैषादीकपालकम् ॥९॥

8. कपालगानफलम्

इति सप्त कपालानि ग न्यग्ह्योदितैः पदैः ।
स्वरंध्र पार्वतीकान्तस्तुतौ कल्याणभाग्भवेत् ॥१०॥

321

Kambala Songs

(ख) कम्बलगानम्

यत्र ग्रहो ऽंशो ऽपन्यासः पञ्चमो बहुलस्तु रिः ।
सो ग्यासो मध्यगान्धारास्त्वल्पास्तत्कम्बलं मतम् ॥११॥

पञ्चमीजातिसञ्जातमल्पताबहुतावशात् ।
स्वराणां बहवो भेदास्तस्य पूर्वैरुदीरिताः ॥१२॥

प्रीतः कम्बलगानेन कम्बलाय वरं ददौ ।
पुरा पुरारिरिच्छापि श्रीयते तैरतः शिवः ॥१३॥

(ग) कपालगतपदानि

कपालानां क्रमाद् भूमौ ब्रह्मप्रोक्तां पदावलीम् ।
झण्टुं झण्टुं ॥१॥   खट्वाङ्गधरं ॥२॥   दंष्ट्राकरालं ॥३॥

Advanced Section – Jati Formations

In this section for the "Sadhak" and student of Indian Swara Shashtra, examples of Jatis as laid out by Sharangdev are listed. Jatis are the mother of all Ragas in essence and hence practicing and being familiar with them makes one appreciate intimately the Ragas that evolved from them. This advanced section is for a student who is already familiar with Swara Shashtra to some degree and is practicing singing or melody creation via an instrument. The examples are in Sanskrit and Devanagari lipi (Hindi language), hence we assume this prerequisite knowledge.

We give these Jatis here so that they can be applied in practice to improve one's understanding of many different Swara and Melody combinations used in composing melodic structures. Translating these in the English language would be essentially counterproductive, as late Dr. R.K. Shringy in his 1977 volume in English has done so already for the seeker who cannot read Hindi. The advanced section is for those willing to put the science and theory in practice and continue their progress with the ideas described here.

18 Jatis - Examples

Shadji

ग. जातीनां विशेषलक्षणानि

अथ प्रत्येकमेतासां जातीनां लक्ष्म कथ्यते ॥५९॥

1. षाड्जी जातिः

षाड्ज्यामंशाः स्वराः पञ्च निषादर्षभवर्जिताः ।
निलोपात्याडवं सो ऽत्र पूर्णत्वे काकली क्वचित् ॥६०॥
सगयोः सधयोश्चात्र संगतिर्बहुलस्तु गः ।
गान्धारे ऽंशे न नेलोंपो मूर्च्छना धैवतादिका ॥६१॥
त्रिधा तालः पञ्चपाणिरत्र चैककलाऽऽदिकः ।
क्रमान्मार्गाश्चित्रवृत्तिदक्षिणा, गीतयः पुनः ॥६२॥

मागधी संभाविता च पृथुलेति क्रमादिमाः ।
नैष्क्रामिकध्रुवायां च प्रथमे प्रेक्षणे स्मृतः ॥६३॥

विनियोगे, द्वादशात्र कला अष्टलघुः कला ।

[तत्र साकल्येन पदयोजना]

[तं भवललाटनयनाम्बुजाधिकं नगसुनुप्रणयकेलिसमुद्भवम् ।
सरसकृततिलय-पङ्कानुलेपनं प्रणमामि कामदेहेन्धनानलम् ॥]

324

अस्यां षाड्ज्यां षड्जो न्यासः । गान्धारपञ्चमाबपन्यासौ ।
वराटी दृश्यते । अस्याः प्रस्तारः:—

१. षाड्जी

१. सा सा सा सा पा निध पा धनि
   तं भ ब ल   ला   ट

२. रो गम गा गा सा रिग धल धा
   न य नां   बु जा   धि

३. रिग सा रो गा सा सा सा सा
   कं

४. धा धा नी निसं निध पा सां सां
   न ग सू   नु प्र ण य

५. नो धा पा धनि रो गा सा गा
   के   लि   स मु   द्भ

६. सा धां धंनि पां सा सा सा सा
   बं

७. सा सा गा सा मा पा मा मा
   स र स कृ त ति ल क

८. सा गा मा धनि निध पा गा रिग
   पं   का नु ले प

९. गा गा गा गा सा सा सा सा
   नं

१०. धां सा री गरि सा मा मा मा
   प्र ण मा   मि का   म

११. धा नी पा धनि री गा री सा
   दे   हूँ   ध ना न

१२. रिग सा री गा सा सा सा सा
   लं

Arshbhi

2. आर्षभी जातिः

आर्षभ्यां तु त्रयो ऽंशाः स्युर्निषादर्षभधैवताः ॥६४॥
द्विश्रुत्योः संगतिः शेषलंङ्घनं पञ्चमस्य च ।
धाडवं षड् ज्जलोपेन सपलोपादिहौडवम् ॥६५॥
मूर्च्छना पञ्चमादिश्च तालश्चच्चत्पुटो मतः ।
अष्टौ कला भवन्तीह विनियोगस्तु पूर्ववत् ॥६६॥

अस्यामार्षभ्यामृषभो न्यासः । अंशा एवापन्यासाः । देशोमधुकर्यौ
दृश्येते । अस्याः प्रस्तारः:—

२. आर्षभी

१. री गा सा रिग मा रिम गा रिरि
   गु ण लो   च ना   धि

२. री री निध निध गा रिम मा पनि
   क म न   न्त म म र

३. मा धा नी धा पा पा सा गा
   म ज र म   क्ष य

४. नी धनि री गरि सर्धं गरि री री
   म जे   यं

५. री मा गरि सर्धं सस रि रिग मम
   प्र ण   मा   मि दिव्य

६. निध पा री री रिय गरि सर्धं सा
   म णि द   पं णा   म

७. रिस रिस रिग रिग मा मा मा गरि
   ल नि के   तं

८. पा नि री मा गरि सर्धं गरि गरि
   भ व म मे   यं

[तत्र साकल्येन पदयोजना]

[गुणलोचनाधिकमनन्तममरमजरमक्षयमजेयम् ।
प्रणमामि दिव्यमणिदर्पणामलनिकेतं भवममेयम् ॥]

Gandhari

3. गान्धारी जातिः

पञ्चांशा रिधवर्ज्याः स्युर्गान्धार्यां संगतिः पुनः ।
न्यासांशाभ्यां तदन्येषां धंबतादृषभं व्रजेत् ॥६७॥

रिलोपरिधलोपाभ्यां षाडबौडुबिते क्रमात् ।
पञ्चमः षाडबढैवी निसमध्यमपञ्चमाः ॥६८॥

अंशा द्विषन्त्यौडुबितं कलाः षोडश कीर्तिताः ।
मूर्च्छना धंबतादिः स्यात्तालश्चचलपुटो मतः ॥६९॥

विनियोगो ध्रुबागाने तृतीयप्रेक्षणे भवेत् ।

अस्यां गान्धार्यां गान्धारो न्यासः । पड्जपञ्चमाबपन्यासी ।
गान्धारपञ्चमदेशोवेलावत्यो दृश्यन्ते । अस्याः प्रस्तारः:—

३. गान्धारी

१. गा गा सा नीं सा गा गा गा
    ए   तं

२. गा गम पा पा धप मा निध निसं
    र ज नि ब धू   मु ख

३. निध पनि मा मपरि गा गा गा गा
    बि   भ्र म   दं

४. गा गम पा पा धप मा निध निसं
    नि शा म य ब रो   ह

५. निध पनि मा मपरि मा गा मा सा
    त व मु ख बि ला   स

६. गा सा गा गा गा गम गा गा
    ब पु श्रा ह   म म ल

७. गा गम पा पा धप मा निध निसं
    मृ दु कि र ण

८. निध पनि मा म॑परि गा गा गा गा
   म   मृ   त भ   बं

९. री गा मा पध री गा सा सा
   र ज त गि रि शि ख र

१०. नीं नीं नीं नीं नीं नीं नीं नीं
    म   णि श क ल शं   ख

११. गा गम पा पा धप मा निध निसं
    ब र   यु ब ति दं   त

१२. निध पनि मा म॑परि गा गा मा गा
    पं   त्ति नि   भं

१३. नी नो पा नो गा मा गा सा
    प्र ण मा   मि प्र ण य

१४. गा सा गा गा गा गम गा गा
    र ति क ल ह र ब नु

१५. गा पा मा मा निध निसं निध पनि
    दं

१६. मा प॑रिग गा गा गा गा गा गा
    श   शि   नं

[तत्र साकल्येन पदयोजना]

[एतं रजनिबधूमुखविभ्रमदं निशामय बरो॑ह
   भवमुखविलासबपुश्रा॑हमसलमृदुकिरणममृतभवम् ।
रजतगिरिशिखरमणिशकलशङ्खबरयुवतिदन्तपड् क्तिनिभं
   प्रणमामि प्रणयरतिकलहरवनुबं शशिनम् ॥]

328

Madhyama

4. मध्यमा जातिः

पञ्चांशा मध्यमायां स्युरगान्धारनिषादकाः ॥७०॥

षड्जमध्यमबाहुल्यं गान्धारो ऽल्पो ऽत्र षाडवम् ।
गलोपान्निगलोपेन त्वौडुवं स्यात्कलाऽष्टकम् ॥७१॥

ऋष्यभादिर्मूर्च्छना स्यात्तालऋचतुष्टो मतः ।
विनियोगो ध्रुवागाने द्वितीयप्रेक्षणे भवेत् ॥७२॥

अस्यां मध्यमायां मध्यमो न्यासः । अंशा एवापन्यासाः ।
चोक्षषाडवदेशगान्धाल्यो दृश्यन्ते । अस्याः प्रस्तारः—

४. मध्यमा

१. मा मा मा मा पा धनि नी धप
   पा   तु भ ब सू

२. मा पम मा सा मा गा री री
   धं जा   न न

३. पा मा रिम पम मा मा मा मा
   कि रो ट

४. मां निध निसं निध पम पध मा मा
   म णि ब   पं   णं

५. नीं नीं री री नीं री री   पा
   गौ   री   क र प

६. नीं मप मा मा सा सा सा सा
   ल्ल वां   गु लि सु

७. गं नी सां गां धप मा धनि सां
   ते   जि तं

८. पा सां पा निधप मा मा मा मा
   सु किर   णं

[तत्र साकल्येन पदयोजना]

[पातु भवमूर्धंजाननकिरीटमणिदर्पणम् ।
गौरीकरपल्लवाङ्गुलिसुतेजितं सुकिरणम् ॥४॥]

329

Panchami

5. पञ्चमी जातिः

रिपाबंशौ तु पञ्चम्यां सगमाः स्वल्पका मताः ।
रिमयोः संगतिगँच्छेन्पूर्णत्वे गान्त्रिषाद्वकम् ॥७३॥

क्रमाद्गेन निगाम्यां च षाडबौद्धवता मता ।
ऋषभो ऽशस्त्यौद्धवितं द्वेष्टपष्टौ च कला मताः ॥७४॥

मूर्च्छनाऽवि तु पूर्वाबत्प्रेक्षणं तु तृतीयकम् ।

अस्यां पञ्चम्यां पञ्चमो न्यासः । ऋषभपञ्चमनिषादा
अपन्पासाः । चोक्षपञ्चमदेशयान्धाल्यो दृश्यन्ते । अस्याः प्रस्तारः:—

५. पञ्चमी

१. पा धनि नी नी मा नी मा पा
   ह रं म्रू र्ध जा न

२. गा गा सा सा मां मां पां पां
   नं म हे श म म र

३. पां पां धां नीं नीं नी गा सा
   प ति बा हु स्तं भ

४. पा मा धा नी निध पा पा पा
   न म नं तं

५. पा पा री' री' री' री' री' री'
   प्र ण मा मि पु रु ष

६. मां निग सा सध नी नीं नी नी
   मु ख प प ल क्ष्मी

७. सां सां सां मा पा पा पा पा
   ह र मं बि का प

८. धा मा धा नी पा पा पा पा
   ति म जे यं

[तत्र साकल्येन पदयोजना]

[हरं मूर्धजाननं महेशममरपतिबाहुस्तम्भनमनन्तम् ।
तं प्रणमामि पुरुषमुखपपलक्ष्मीहरमम्बिकापतिमजेयम् ॥]

330

Dhaivati

6. धैवती जातिः

स्तो धैवत्यां रिधाबंशौ लङ्घ्यघाबारोहिणौ सपौ ॥७५॥

पलोपात्खाडबं प्रोक्तमौडुवं सपलोपतः ।
ऋषभादिर्मूर्च्छना स्यात्तालो मार्गश्च गीतयः ॥७६॥

विनियोगश्च षाड्जीवत्कला द्वादश कीर्तिताः ।

अस्यां धैवत्यां धैवतो न्यासः । ऋषभमध्यमधैवता अपन्यासाः ।
बौक्षकैशिकदेशीसिंहल्यो दृश्यन्ते । अस्याः प्रस्तार :--

६. धैवती

१.   धा धा निध पध मा मा मा मा
     त   ह   णा   म   लँ   बु

२.   धा धा निध निसं सां सां सां सां
     म   णि भू   वि ता   म

३.   सध धा पा मध धा निध धनि धा
     ल   शि रो   जं

४.   सा सा रिग रिग सा रिग सा सा
     भु ज   गा   धि पै   क

५. धां धां नीं पां धां पां मां मां
   कुं   ड ल वि ला   स

६. धां धां पां मंघं धां निघं धंनि धां
   कु त शो   भं

७. धा धा निसं निसं निध पा पा पा
   न ग सू   नु ल   क्ष्मी

८. रिग सा सा सा नीं नीं नीं नीं
   बे हा   र्घ मि   धि

९. सा रिग रिग सा नीं सा धां धां
   त श रो   रं

१०. रीं गंरि संगं मां मां मां मां मां
    प्र ण मा   मि भू   त

११. नी नी धा धा पा रिग सा रिग
    मी   तो प हा   र

१२. पा धा सा मा धा नो धा धा
    प रि तु   ष्टं

[तत्र साकल्येन पदयोजना]

[तरुणामलेन्दुमणिभूषितामलशिरोजं
    भुजगाधिपंककुण्डलविलासकृतशोभम् ।
नगसुनुलक्ष्मीदेहार्धमिश्रितशरीरं
    प्रणमामि भूतगीतोपहारपरितुष्टम् ॥]

Naishadi

7. नैषादी जातिः

नैषाद्यां निरिगा अंशा अनंशाबहुलाः स्मृताः ॥७७॥

षाडवौडुवलङ्ध्याः स्युः पूर्वावद्धिनियोजनम् ।

चञ्चत्पुटः षोडशात्र कला गादिश्च मूर्च्छना ॥७८॥

अस्यां नैषाद्यां निषादो ग्यासः । अंशा एवापन्यासाः । चोक्ष-
साधारितदेशीवेलाबल्यो दृश्यन्ते । अस्याः प्रस्तारः:—

७. नैषादी

१. नी नो नो नी सां धा नी नी
   तं   सु र वं   दि त

२. पा मा सा धां नीं नीं नीं नीं
   म हि य म हा   सु र

३. सा सा गा गा नी नी धा नी
   म थ न मु मा   प ति

४. सां सां धा नी नी नी नी नी
   भो   ग यु तं

५. सा सा गा गा मां मां मां मां
   न ग सु त का   मि नी

६. नीं पां धां पां मां मां मां मां
   दि   व्य बि शे   ष क

७. री॑ गां सं॑ सं॑ री॑ गां नी नी
   सू   च क शु भ न ख

८. नी नी पा धनि नी नी नी नी
   द   पं ण   कं

९. सा सा गा सा मा मा मा मा
   अ हि मु ख म णि ख चि

१०. मां मां मां मां नीं धां मां मां
   तो   ज्ज्व ल नू   पु र

११. धा धा नी नो री गा मां मां
   बा ल   भु जं ग   म

१२. मां मां पां धां नीं नीं नीं नीं
   र व क लि   तं

१३. पां पां नीं नीं री री री री
   द्रु त म भि व्र जा   मि

१४. री मा मा मा री गा सा सा
   श र ण म निं   दि त

१५. धा मा री गा सा धा नो नी
   पा   द यु ग पं   क

१६. पां मां री॑ गां नी नी नी नी
   ज बि ला   सं

Shadaja Kaishiki

8. षड्जकैशिकी जातिः

अंशाः स्युः षड्जकैशिक्यां षड्जगान्धारपञ्चमाः ।
ऋषभे मध्यमे ऽल्पत्वं धनिषादौ मनाग्बहू ॥७९॥

चञ्चत्पुटः षोडशास्यां कलाः स्युर्विनियोजनम् ।
प्रावेशिक्यां ध्रुवायां स्वात्प्रवेशणे तु द्वितीयके ॥८०॥

अस्यां षड्जकैशिक्यां गान्धारो न्यासः । षड्जनिषादपञ्चमा
अपन्यासाः । प्रागुक्ता गान्धारपञ्चमहिन्दोलकदेशीवेलावल्यो दृश्यन्ते ।
अस्याः प्रस्तारः—

८. षड्जकैशिकी

१. सा सा मां   पां गरि मग मा मा
    वे

२. मा मा मा मा सां सां सां सां
    वं

३. धा धा पा पा धा धा री रिम
    अ स क ल श शि ति ल

४. री री नीं नीं नीं नीं नीं नीं
    कं

335

५. धा धा पा धनि मा मा पा पा
   द्धि, र द ग ति

६. धा धा पा धनि धा धा पा पा
   नि पु ण म ति

७. सा सा सा सा सा सा सा सा
   मु ग्ध मु खां बु

८. धा धा पा धा धनि धा धा धा
   र ह दि ब्य कां ति

९. सा सा सा रिग सा रिग धा धा
   ह र मं बु बो व

१०. मा धा पा पा धा धा नी नी
   धि नि ना दं

११. री री गा सा सां सां सां गां
   अ च ल व र सू नु

१२. धां रिसं रीं संरि रीं सां सां सां
   वे हा धं मि भि

१३. सा सरि री सरि री सा सा सा
   त श री रं

१४. मा मा मा मा निध पध मा मा
   प्र ण मा मि तम हं

१५. नी नी पा पम पा पम पध रिग
   अ नु प म मु ळ क म

१६. गा गा गा गा गा गा गा गा
   लं

Shadajotyava

9. षड्जोदीच्यवा जातिः

अंशाः समनिधाः षड्जोदीच्यवायां प्रकीर्तिताः ।
मिथश्च संगतास्ते स्युर्मन्द्रगान्धारसूरिता ॥८१॥
षड्जर्षभौ भूरितारौ रिलोपात्खाडवं मतम् ।
औडुवं रिपलोपेन धैवते ऽ शे न षाडवम् ॥८२॥

षड्जोबवृगीतितालादि गान्धाराविश्च मूर्छना ।
द्वितीये प्रेक्षणे गाने ध्रुवायां विनियोजनम् ॥८३॥

अस्यां षड्जोदीच्यवायां मध्यमो न्यासः । षड्जर्षभधैवताचपन्यासौ ।

अस्याः प्रस्तारः—

१.	षड्जोदीच्यवा

१. सा सा सा सा सां मां गां गां
   शै   ले

२. गा मा पा मा गा मा मा धा
   श   सू   तु

३. सा सा मा गा पा पा नी धा
   शै   ले   श सू   तु

४. धा नी सा सा धा नी पा मा
   प्र ण य   प्र सं   ग

५. गां सा सा सा सा सा सा गां
   स बि ला   स खे   ल

६. धा धा पा धा पा नी धा धा
   न बि नो   वं

७. सा गां गां गां गां गां सा सा
   अ   धि क

८. नी धा पा धा पा धा धा धा
   मु   खें   दु

९. सां सां मा गा पा पा नी धा
   अ धि क   मु खें   दु

१०. धा नी सां सां धा नी पा मा
    न य नं   न मा   मि

११. गां सा सा सा सा सा सा गां
    दे   बा   सु रे   श

१२. धा धा पा धा मां मां मां मां
    त व ह चि रं

शैलेऽक्षराभ्यां प्रथमा द्वितीया तु शशसूनुना ।
तैः पञ्चभिस्तृतीया स्यात्सप्तमी स्वधिकाक्षरैः ॥८४॥
मुखेन्दुना ऽष्टमी त्वस्यां षड्भिस्तनंबमी कला ।

[तत्र साकल्येन पदयोजना]

[शैलेशसूनुप्रणयप्रसङ्गसबिलासखेलनबिनोदम् ।
अधिकमुखेन्दुनयनं नमामि देवासुरेश तब रुचिरम् ॥]

Shadaj Madhyam

10. षड्जमध्यमा जातिः

अंशाः सप्त स्वराः षड्जमध्यमायां मिथश्च ते ॥८५॥
संगच्छन्ते निरल्पो ऽऽषाद्गादृते वाचितां विना ।
निलोपनिगलोपाभ्यां षाडवौडुविते मते ॥८६॥
षाडवौडुवयोः स्यातां द्विश्रुती तु विरोधिनौ ।
गीतितालकलाऽऽदीनि षाड्जीवन्मूर्च्छना पुनः ॥८७॥
मध्यमादिरिह ज्ञेया पूर्वबद्धिनियोजनम् ।

अस्यां षड्जमध्यमायां षड्जमध्यमौ न्यासौ । सप्त स्वरा
अपन्यासाः । अस्याः प्रस्तारः:—

१०. षड्जमध्यमा

१. मा गा सग पा धप मा निध निम
   र ज नि ब धू   मु ष

२. मां मां सां रिंगं संगं निध पध पा
   वि ला   स लो   च

३. मा गा री गा मा मा सा सा
   नं

४. मा मगम मा मा निध पध पम गमम
   प्र वि क सि त कु मु व

५. धा पध परि रिग मग रिग सधस सा
   व ल फे न सं   नि

६. निध सा री मगम मा मा मा मा
   भं

७. मां मां मंगंमं मंचं धंपं पंधं पंमं गंमंगं
   का   मि ज न न य न

८. धा पध परि रिग मग रिग सधस ता
   हृ द या भि नं   वि

९. मा मा धनि धस धप मप पा पा
   नं

१०. मां मंगंमं मां निधं पंधं पंमंगं गां मां
    प्र ण मा   मि दे वं

११. धा पध परि रिग मग रिग सधस सा
    कु मु वा धि वा   सि

१२. निध सा री मगम मा मा मा मा
    नं

[तत्र साकल्येन पदयोजना]

[रजनिवधूमुखविलासलोचनं
    प्रविकसितकुमुवदलफेनसंनिभम् ।
कामिज्ञनननयनहृदयाभिनन्विनं
    प्रणमामि देवं कुमुवाधिवासिनम् ॥]

339

Gandharoditchyava

११. गान्धारोद्वीच्यवा जातिः

गान्धारोद्वीच्यवायां तु द्वावंशो षड्जमध्यमौ ॥८८॥
रिलोपात्त्वाडवं ज्ञेयं पूर्णत्वे ड'शेतराल्पता ।
अल्पा निधपगान्धाराः वाडवत्वे प्रकीर्तिताः ॥८९॥
रिधयोः संगतिर्ज्ञेया पंचतादिभ्र मूर्च्छना ।
तालभ्रञ्चत्युटो ज्ञेयः कलाः षोऽश कीर्तिताः ॥९०॥
विनियोगो भ्रूवागाने चतुर्थप्रेक्षणे मतः ।

अस्यां गान्धारोद्वीच्यवायां मध्यमो न्यासः । षड्जर्षभतादव-
पन्यासौ । अस्याः प्रस्तारः:—

११. गान्धारोद्वीच्यवा

१. सा सा पा मा पा धप पा मा
    सौ

२. धा पा मा मा सा सा सा सा
    म्य

३. धा नी सा सा मा मा पा पा
    मौ   री   मु खां   बु

४. नी नी नी नो नी नी नी नी
    उ ह वि   व्य ति ल क

५. मा मा धा निस नी नी नी नी
    प रि चुं   बि ता   चि

६. मा पा मा परिग गा गा सा सा
    त सु पा   वं

७. गा मग पा पध मा धनि पा पा
    प्र बि क सि त है   म

८. री गा सा सध नो नी धा धा
    क म ल नि भं

९. गा रिग सा सनि गा रिग सा सा
    अ ति ए चि र कां   ति

१०. सा सा सा मा मनि धनि नी नी
    न ख द   र्प णा   म

११. सां पां मां पंरिगं गां गां सां सां
    ल नि के   तं

१२. गां सां गां सां मां पां मां पंरिगं
    म न सि ज श री र

१३. गां मां गां सां गां गां गां सां
    ता   ड नं

१४. नीं' नीं' पां धां नीं' गां गां गां
    प्र ण मा   मि गौ   री

१५. नीं' नीं' धां पां धां पां मां पां
    च र ण यु ग म न प

१६. धां पां सां सां मां मां मां मां
    मं

[तत्र साकल्येन पदयोजना]

[सौम्यगौरीमुखाम्बुरुहदिव्यतिलकपरिचुम्बितार्चितसुपादं
प्रविकसितहेमकमलनिभम् ।

अतिरुचिरकान्तिनखवर्पणामलनिकेतं
मनसिजशरीरताडनं प्रणमामि गौरीचरणयुगमनुपमम् ॥]

Raktgandhari

12. रक्तगान्धारी जातिः

अंशाः स्यू रक्तगान्धार्यां पञ्च धर्षभर्वजिताः ॥९१॥
रिमतिक्रम्य सगयोः कार्ये संनिधिमेलने ।
रिलोपरिधलोपाभ्यां षाडवौडुवमिष्यते ॥९२॥

बहुत्वं निधयोरंशः पञ्चमो द्रेष्टि ।डवम् ।
द्विषन्त्यौडुवितं षड्जनिमपाः संगती सगौ ॥९३॥

पञ्चपाण्यादि पाड्जीवद्दृषभादिस्तु मूर्छना ।
तृतीयप्रेक्षणगतश्र् वायां विनियोजनम् ॥९४॥

अस्यां रक्तगान्धार्यां गान्धारो न्यासः । मध्यमोऽपन्यासः । अस्याः
प्रस्तारः...

१२. रक्तगान्धारी

१. पा नी सा सा गा सा पा नी
   तं   बा   ल र ज नि

२. सां सां पा पा मा मा गा गा
   क र ति ल क भू   ष

३. मा पा धा पा मा पा धप मग
   ण वि भू

४. मा मा मा मा मा मा मा मा
   ति

341

५. धॊ नींॅ पां संपं धां नींॅ पां पां
   ०

६. मां पां मों धंति पां पां पां पां
   ०

७. री गा मा पा पा पा मा पा
   प्र ण मा   मि गौ   री

८. री गां मां पां पां पां मां पां
   ब द ना   र बि

९. पा पा पा पा शा पा पा पा
   द

१०. री गा सा सा री गा गा मा
    प्री   ति क रं

११. गां गा पां धंसं धां निंधं पां पां
    ०

१२. मां पां मां पंरिगं गां गं गां गां
    ०

[तच्च साकल्येन पदयोजना]

[तं बालरजनिकरतिलकभूषणबिभूतिम् ।
प्रणमामि गौरीबदनारविन्दप्रीतिकरम् ॥]

Kaishiki

13. कैशिकी जातिः

कैशिक्यामृषभान्ये ऽंशा निधावंशौ यदा तदा ।
न्यासः पञ्चम एव स्यादन्यदा द्विश्रुती मतौ ॥९५॥

अन्ये तु निगपान्न्यासानिधयोरंशयोर्बिंदुः ।
रिलोपरिधलोपेन षाडबौडुवितं मतम् ॥९६॥

रिरल्पो निपबाहुल्यमंशानां संगतिर्मिथः ।
षाडबौडुविते द्विष्ठः क्रमात्पञ्चमधैवतौ ॥९७॥

षाडुजीवत्पञ्चपाण्यादि गान्धारादिस्तु मूर्च्छना ।
पञ्चमप्रेक्षणगतश्च वायां विनियोजनम् ॥९८॥

342

अस्यां कंशिक्यां गान्धारपञ्चमनिषादा न्यासाः । रिवर्ज्याः षट्
सप्त वा स्वरा अपन्यासाः । अस्याः प्रस्तारः—

१२. कंशिकी

१. पा धनि पा धनि गा गा गा गा
के   ली   ह   त

२. पा पा मा निध निध पा पा पा
का म त नु

३. धा नी सां सा सा री री री री
बि भ्र म बि ला   सं

४. सा सा सा री गा मा मा मा
ति ल क यु तं

५. मां धां नीं धां मां धां मां पां
मू धॉं र्ध्व बा   ल

६. गा री सा धनि री री री री
सो म नि भं

७. गा री सा सा था धा मा मा
सु ख क म लं

८. गा गा गा मा मा निधनि नी नी
अ स म   हा   ट

९. गा पा नी नी गा गा गा गा
क स रो जं

१०. गां गां नों नों नीं धं पां पां पां
हृ दि सु ख बं

११. मां पां मां पां पां पां मां मां
प्र ण मा मि लो च

१२. सां मां गां निधनि नीं नीं मां गां
न बि शे   षं

[तत्र साकल्येन पदयोजना]

**[कलीहृतकामतनुबिभ्रमबिलासं तिलकयुतं मूर्धोर्ध्वंबालसोमनिभम् ।
मुखकमलमसमहाटकसरोजं हृदि सुखदं प्रणमामि लोचनविशेषम् ॥]**

343

Madhyomidtchyava

14. मध्यमोदीच्यवा जातिः

पञ्चमांशा सदा पूर्णा मध्यमोदीच्यवा मता ।
लक्ष्म शेषं विजानीयाद् गान्धारोदीच्यवागतम् ॥९९॥
मूर्च्छना मध्यमादिः स्यात्तालश्चञ्चत्पुटो मतः ।
चतुर्थस्य प्रेक्षणस्य ध्रुवायां विनियोजनम् ॥१००॥
अस्यां मध्यमोदीच्यवायां मध्यमो न्यासः । अस्याः प्रस्तारः—

१४. मध्यमोदीच्यवा

१.	पा	धनि	नी	नी	मा	पा		नी	पा
	दे		हा		धं	रू			प
२.	री	री		री	गा	सा	रिग	मा	गा
	म	ति		कां		ति	म		म	ल
३.	नी	नी		नी	नी	नी	नी		नी	नी
	म	म		लें		दु	कं			द
४.	नी	नी		धप	मा	निध	निध		पा	पा
	कु	मु		द	नि	भं			
५.	पा	पा		री	री	री	री		री	री
	चा			मी		क	रां			बु
६.	मा	रिग	सा	सधं	नीं	नीं		नीं	नीं
	ह	ह	दि			व्य		कां	ति

७. मा पा नी सा पा पा गा गा
   प्र ब र ग ण पू   जि

८. गा पां मां निधं नीं नीं सा सा
   त म जे   यं

९. पां पां मां धंनि पां पां पां पां
   सु रा भिष्टु त म निल

१०. मां पां मां रिग गा मा गा गा
   म नो ज   व   मं बु

११. गा पा मा पा नी नी नी नी
   बो   व धि नि ना   व

१२. मा पा मा परिग गा गा गा गा
   म ति हा   सं

१३. गां गां गां गां मां निधं नी' नी'
   शि वं शां त म सु र

१४. नी नी धप मा निध निध पा पा
   च म्भू म य नं

१५. री' गां सां सां मां निधंनिं नी' नी'
   वं दे त्रै लो क्य

१६. नी' नी' धं पा धं पां मां मां
   न त च र णं

[तत्र साकल्येन पदयोजना]

[देहार्धरूपमतिकान्तिममलममलेन्दुशुन्दकुमुदनिभं
   स्वामीकराम्बुरुहदिव्यकान्तिप्रबरगणपूजितमजेयम् ।
सुराभिष्टुतमनिलमनोजवमम्बुबोधिनिनादमतिहासं
   शिवं शान्तमसुरचभूमयनं बन्दे त्रैलोक्यनतचरणम् ।।]

Karmaravi

15. कार्मारवी जातिः

कार्मारव्यां भवन्त्यंशा निषादरिपधैवताः ।
बहुबीञ्तरसार्गतत्वादवंशाः   परिकीर्तिताः ।।१०१।।
षान्धारोऽल्यन्तबहुलः सर्वांशस्वरसंगतिः ।
षण्जलयुदः षोडशान्न कलाः, षड्जादिमूर्च्छना ।।१०२।।
पञ्चमस्यास्य प्रेक्षणस्य ध्रुवायां विनियोजनम् ।

अस्यां कार्मारव्यां पञ्चमो न्यासः । अंशा एवापन्यासाः । अस्याः
प्रस्तारः—

345

१५. कार्मारबी

१. री री री री री री री री
    तं   स्था   णु ल लित

२. मा गा सा गा सा नी नी नी
    बा   मां   ग स   त्त

३. नीं मां नीं मां पां पां गा गा
    म ति ते   जः प्र स र

४. गा पा सा पा नी नी नी नी
    सौ   धां   शु कां   ति

५. रों गां सां नीं रों' गां रों' मां
    फ णि प ति सु खं

६. री गा री सा नो धनि पा पा
    उ रो बि पु ल सा   ग

७. मां पा मां तंरिबं गा गा गा गा
    र नि के   तं

८. री रो गा सम मा मा पा पा
    सि त पं   न गॅ   द्र

९. मा पा मा परिग ग गा गा गा
    म ति कां   तं

१०. धा नी पा मा धा नो सा सा
     ष   ण्मु ख बि नो   द

११. मी नी नो मी नी नी नी नी
     क र प   ह्ल बां   गु

१२. मां मां धां नीं सनिनि धा पा पा
     लि बि ला   स   की   न

१३. मा पा मा परिग गा गा गा गा
न बि नो   बं

१४. नी नी पा धनि गा गा गा गा
प्र ण मा   मि दे   व

१५. सां रों गां सां नीं नीं नीं नीं
य   ज्ञो   प   बी   त

१५. नीं नीं धां धां पां पां पां पां
कं

[तत्र साकल्येन पदयोजना]

[तं स्वाणुर्लालतवामाङ्गसक्तमतितेजः-प्रसरसौधांगुकान्ति-फणिपति-
मुखमुरोविपुलसागरनिकेतं सितपन्नगेन्द्रमतिकान्तम् ।
धप्मुखविनोबकरपल्लवाङ्गुलिबिलासकीलनविनोदं
प्रणमामि देवयज्ञोपवीतकम् ॥]

Gandharpanchami

16. गान्धारपञ्चमी जातिः

अंशो गान्धारपञ्चम्यां पञ्चमः, संगतिः पुनः ॥१०३॥
कर्तव्या ज्ञापि गान्धारीपञ्चम्योरिव भूरिभिः ।
चञ्चत्पुटः षोडशात्र कला गाविश्र मूर्छना ॥१०४॥
तुर्यप्रेक्षणसम्बन्धिश्र वारगाने   नियोजनम् ।
अस्यां गान्धारपञ्चम्यां गान्धारो न्यासः । ऋषभपञ्चमावपन्यासौ ।
अस्याः प्रस्तारः—

१६. गान्धारपञ्चमी

१. पा मप मध नी धप मा धा नी
कां

२. सनिनि धा पा पा पा पा पा पा
तं

३. धा नी सा सा मा मा पा पा
    वा   में   क दे   श

४. नी नी नी नी नी नी नी नी
    त्रॅं   खो   ल मा   न

५. नी नी धप मा निध निध पा पा
    क म ल नि भं

६. पा पा री री री री री री
    व र सु र भि कु सु म

७. मा रिग सा सध नी नी नी नी
    गं   धा   धि वा   सि

८. नी नी सां रिंसं रीं रीं रीं रीं
    त म नो   ज्ञ

९. नो गा सा निग सा नीं नीं   नीं
    न ग रा   ज सु   नु

१०. नीं मां नीं मां पां पां गा गा
    र ति रा   ग र भ स

११. गा पां मां पां नीं नीं नीं नीं
    के   ली   कु च   प्र

१२. मा पा मा परिग गा गा गा गा
    हृ ली लं   तं

१३. नीं नीं पां धां नीं गा गा गा
    प्र ण मा   मि दे   वं

१४. नीं मीं नीं नीं नीं नीं नीं नीं
    चं   द्रा   र्धं मं   डि

१५. मां मां धां नीं सनिनि धा पा पा
    त वि ला   सकी लीं

१६. मा पा मा परिग गा गा गा गा
    न वि नो   वं

[तत्र साकल्येन पदयोजना]

[कान्तं वामकुदेशप्रेह्लोलमानकमलनिभं वरसुरभिकुसुमगन्धाधि-
वासितमनोज्ञनगराजसूनुरतिरागरभसकेलीकुचप्रहृलीलम् ।
तं प्रणमामि देवं चन्द्रार्धमण्डितविलासकीलनविनोदम् ॥]

Andhri

17. आन्ध्री जातिः

आन्ध्रप्रधामंशा निरिगपवा रिगपोनिधवोस्तथा ॥१०५॥

संगतिःर्वासपर्यन्तमंशानुक्रमतो   व्रजेत् ।

षाडवं षड्जलोपेन मध्यमादिस्तु   भूर्च्छना ॥१०६॥

पूर्वावत्तु   कलातालविनियोगः   प्रकीतितः।

१७. आन्ध्री

१. गा री री री री री री री
    त ह णें   डु कु सु म

२. री गा री गा री री री री
    ख चि त ज टं

३. री री गा गा री री भा मा
    त्रि दि व न बी स लि ल

४. री गा सा धनि नीं नीं नीं नीं
    धौ   त मु   खं

५. नीं री नीं रीं र्धनि र्धनि पां पां
    न म सू   नु प्र ण यं

६. मां पां मां रिग गा गा गा गा
    बे   द नि धिं

७. री री गा सस मा मा पा पा
    प रि णा   हि तु हि न

८. मां पां मा रिग गा गा गा गा
    शे   ल गु   हं

९. र्धां नीं गा गा गा गा गा गा
    अ मृ त भ वं

१०. पा पा मा रिग गा गा गा गा
    गु ण र हि तं

११. नी नी नीं नीं री री री री
    त म ब निर बि श शि

१२. री री गा नी सा सा नी नी
    ज्व ल न ज ल प ब न

349

१३. पां पां मां रिंगं गां गां गां गां
     ग   ग न त नुं

१४. रीं रीं गां संमं मां मां पां पां
     श   र णं   व्र जा   मि

१५. मां मां नीं नीं सां रीं गां पां
     शु भ म ति कृ त नि ल

१६. रिंगं गां गां गां गां गां गां गां
     यं

[तत्र साकल्येन पदयोजना]

[तरुणेन्दुकुसुमखचितजटं   त्रिदिवनदीसलिलधौतमुखं
नगसूनुप्रणयं वेदनिधिं परिणाहितुहिनशैलगृहम् ।
अमृतभवं गुणरहितं तमवनिरविशशिज्वलनजलपवनगगनतनुं
शरणं   व्रजामि   शुभमतिकृततनिलयम् ।।]

Nandayanti

18. नन्दयन्ती जातिः

नन्दयन्त्यां पञ्चमो ऽंशो गान्धारस्तु ग्रहः स्मृतः ।।१०७।।

कंश्चित्तु पञ्चमः प्रोक्तो ग्रहो ऽस्यां गीतवेदिभिः ।
मन्द्रर्षभस्य बाहुल्यं षाडवं षड्जलोपतः ।।१०८।।

हृष्पका मूर्च्छना तालः पूर्वावद् द्विगुणाः कलाः ।
विनियोञ्जे ध्रुवागाने प्रथमप्रेक्षणे भवेत् ।।१०९।।

अस्यां नन्दयन्त्यां गान्धारो न्यासः । मध्यमपञ्चमावपन्यासौ ।
अस्याः प्रस्तारः:—

१८. नन्दयन्ती

१. गां गां गां गां पां पां धप मां
   सौ

२. धा धा धा धा धां नीं सनिनि धा
   ०

३. पां पां पां पां पां पां पां पां
   म्यं

४. धां नीं मां पां मां मां मां गां
   बे　　वां　　ग　बे　　द

५. मा रो गा गा गा गा गा गा
   क र क म ल यो　नि

६. मा मा पा पा धा निध पा पा
   त मो र जो बि द

७. धा नी मा पा गा गा गा गा
   जि तं

८. गम　पा पा पा मा मा गा गा
   हरं

९. धा नी मा पा गा गा गा गा
   भ ब ह र क म ल गृ

१०. मा मा मा मा मा मा मा मा
    हुं

११. रो गा मा पा पम पा पा नी
    शि वं शां　तं　सं　नि

१२. रीं रीं रीं रीं पां पां मां मां
    बे　श न म पू　बं

१३. धां नीं सांर्निनि धां पां पां पां पां
    भू ष　　ण ली　लं

१४. धां नीं मां पां गां गां गां गां
    उ र गे　श भो　ग

१५. गा पा पा पा धा मा गा मा
    भा　सु र शु भ पृ थु

१६. धा धा नी धा पा पा पा पा
    लं

१७. री गा मा पा पम पा पा नी
अ च ल प ति सू नु

१८. रीं रीं रीं रीं पां पां पां पां
क र पं क जा म

१९. पा पा पा पा धा मा मा मा
ल बि ला स की ल

२०. नीं पां गां गमं गां गां गां गां
न बि नो दं

२१. रीं रीं गां गां मां मां मां मां
स्फ टि क म णि र ज त

२२. नी पा नी मा नी धा पा पा
सि त न व दु कू ल

२३. सां सां धनि धा पा पा पा पा
क्षी रोद सा ग

२४. मा पा मा परिग गा गा सां सां
र नि का शं

२५. री री गा गा मा मा पा पा
अ ज शि र: क पा ल

२६. री री री गा मा रिग मा मा
पृ थु भा ज नं

२७. मा नी पा नी गा गा मा गा
बं दे सु ख दं

२८. मा मा पा पा धा धनि निध मा
ह र दे ह म म ल

२९. धा धा सा नी धा नी पा पा
म धु सू व न शु

352

३०. री' री' री' री' मा पा धा मा
    ते   जो   धि क   सु

३१. नी नी नी नी धा पा मा मा
    ग ति यो

३२. मा परिग गा गा गा गा गा गा
       नि

[तत्र साकल्येन पदयोजना]

[सौम्यं वेदाङ्गवेदकरकमलयोनि तमोरजोबिर्बजितं हरं
     भवहरकमलगूहं   शिवं   शान्तं   सन्निवेशनमपूर्वं
भूषणलीलमुरगेशाभोगभासुरशुभपृथुलम् ।
अचलपतिसूनुकरपङ्कजामलबिलासकीलनविनोदं
     स्फटिकमणिरजतसितनबदुकूलक्षीरोदसागरनिकाशम् ।
अजशिरःकपालपृथुभाजनं वन्दे सुखवं
     हरवेहममलमधुसूदनसुतेजोऽधिकसुगतियोनिम् ।।]

Epilogue – Swara to 'Raga'

Melody to Composition

As we conclude this analysis of Indian 'Swara Shashtra' we have spanned an entire gamut of thought and feelings from Creator to Creation, From being born out of this Creation to giving birth to our own Melody using our human form as an instrument. As Dr. Rajeshwar Acharya, succinctly points out in his foreword; 'Music' is not created but rather is preexisting and is 'Made Available' in the Universe. What we are doing through our human efforts is to identify the natural science of this process I an effort to extract some notes of melody out of it for our soul and share it with few other souls who can resonate with our 'Swara'. This is the satisfactory objective of this ever-continuing journey.

We have covered the 'Ten' essential Prana of the Indian musical melodic system. We defined the all-permeating universal sound 'Naad' synonymous with the creator himself (Naad Brahman). From this Naad we understood the derivation of two related concepts of Shruti (22 tones) and Swara (7 notes). Then we defined further details of the life force of our Melody by understanding macro melodic groupings of Grama, Murchana-Krama, and Taan.

This then takes us to the practical application of the sounds of Melody described above into further defined tonal structures such as Sadharana, Varnalankara, Jati and Giti. These are the ten essential Prana of Indian Swara Shashtra. These fundamental tonal phenomena constitute the foundation of what we know now as the Indian classical musical system. The Melodic types (Jatis) and their rules of usage (Gitis) lead to the infinite melodic creations and possibilities if our music. The end-result of understanding, studying and applying these ten concepts of Melody in practice is that we obtain our system of structured compositions called Ragas.

Ragas are born of these ten Prana (life force) of 'Swara Shashtra'.

It would be essential in summary here to define the term Raga in more detail. In ancient times, the music derived from the above foundational principles resulted in what was known as 'Jati Gana', which over contemporary times evolved into what we now know as Raga. Thus, Jati Gana is the father of our modern Raga system and familiarizing oneself with the ancient Jati Gana system solidifies our complete evolutionary understanding of the Ragas.

The Sanskrit word 'Raga' derives from the root word 'ranj' that is connected to 'rakti'. 'Rakti' means being absorbed in the pleasure of something that one loves. Moreover, 'ranj' means to color something or to love something with passion. This leads to our evolved meaning of the word 'Raga' that means to be enamored by an object's aesthetic beauty or emotional sense and its sensibilities.

यो'सौ ध्वनिविशेषस्तु स्वरवर्णविभूषित: ।
रञ्जको जनचित्तानां स राग: कथितो बुधै: ॥ *S.R.*, II, p. 3.
'रञ्जनाज्जागते रागो व्युत्पत्ति: समुदाहृता ।
अश्मकर्णादिषट्दक्षौ यौगिको वापि वाचक: ।
योगरूढोऽपया रागो ज्ञेय: पङ्कजशब्दवत् ॥ *Br.D.*, 283, 284.
'रागमार्गस्य यद्रूपं यन्नोक्तनं भरतादिभि: ।
निरूगमे नदरमाभिन्नेथ्यलक्षणसंयुतम् ॥ *Br.D.*, 279.

The musically technical definition of Raga as prescribed by Sharangdev is as follows. "A structured sound formation that employs embellishments (Varnalankara) of musical tones (Shrutis) and special movements of tonal patterns (Jatis) and is delightful to the listener is known as a Raga by the learned and wise."

The technical definition tells us that besides producing the 'rakti' joy of aesthetic pleasure, the idea of 'Raga' over time evolved into a more structured definition since the time of earliest known treatise from Bharata. From the time of Sharangdev in 12ᵗʰ century AD, the Raga system became formally solidified as having a specific technical aspect called 'Lakshya' (characterized music as the objective) and the 'Lakshana' (characteristic features) that define the Raga.

In modern contemporary times, the entire system of Ragas that has been accepted as a base foundation was codified based on the ancient musical treatise by tireless efforts of Pandit V.N. Bhatkhande of Maharashtra in India and this is documented in his textbooks of music. These readily available modern texts cover the entire classification, differentiation and characteristic definitions of each of the many thousands of Ragas that are known. This must be the next springboard for the seeker. We have thus covered in summary the entire 'Science of Melody' and now arrived at the doorsteps of the Raga compositions that are to be felt, experienced and performed in solitude and in front of discerning listeners. The journey, however, does not end there.

What is documented in these Ragas, Jatis are the feelings, and sensibilities of the sages and musicians of past, but what is 'Not Documented' yet are your own creations. The purpose of knowing and understating the roots of our 'Swara Shashtra' is for the soul to dive into the infinite source of creativity in our universe and create ever-new Ragas and musical melodies. This spontaneous creation, when experienced and offered to the 'Divine Providence', is the ultimate objective and source of Joy. Even if one reader of this book shares this emotional feeling and embarks upon that journey of 'Divine Naad', this Authors' soul would be forever satisfied in making this ongoing Naad yatra worthwhile 'Saarthak'.

Bibliography

Some of the ancient musical texts and Philosophical texts referenced by Author in compilation of this volume are as follows;

1. Perspectives on Rasa ("Ras Drishti ki Tarfen me") by Shri Shyam Manohar Goswamyji

2. Sangeet Ratnakar by Sharangdev 4 volume English Translation by Dr. R.K. Shringy, Benaras

3. Sangitanjali Vol 1 to 6 by Pandit Omkarnath Thakur

4. Taal Deepika (4 volumes) by Sw. Shri Mannuji

5. Mridang Ank by Sw. Shri Pagalbaba

6. Nada Rasa by NL Shri Mukundray Goswamy

7. Natya Shastra by Sage Bharat

8. Natya Shastra Tika Commentary by Abhinav Gupt

9. Sangeet Ratnakar by Sarangadev

10. Sangeet Makarand by Sage Narada

11. Shri Subodhiniji Ras Panchadhyayi with Tika from Purushottamji (by Jagganath MoonMoonji Chaturvedi, Chaukhambha Prakasahan Benaras)

12. "Brahma Sutra" by Veda Vyasa and its commentaries by Shri Shyamubava and Shri Hariray Mahaprabhu

13. Discourses on Shri Subodhiniji and Brahmavaad "Laghu Granth Tika Sangrah" by Shri Dixitji Maharaj (father of Shri Shyamubava)

14. Measuring the representational space of music with fMRI: a case study with Sting, Daniel J. Levitin & Scott T. Grafton Pages 548-557 | Received 18 Apr 2016, Accepted 19 Jul 2016, Published online: 12 Aug 2016

15. music on the Mind: an introduction to this special issue of Neurocase, Indre V. Viskontas & Elizabeth Hellmuth Margulis, Pages 484-485 | Published online: 21 Dec 2016

16. Ross, Jessica & Iversen, John & Balasubramaniam, Ramesh. (2016). Motor Simulation Theories of musical Beat Perception... Neurocase. 22.

10.1080/13554794.2016.1242756.

17. https://www.nasa.gov/centers/goddard/universe/black_hole_sound.html

18. https://www.nytimes.com/2003/09/16/science/music-of-the-heavens-turns-out-to-sound-a-lot-like-a-b-flat.html

19. The Implications of Anoraniyan Mahato Mahiyan as related to Vastu Science, 2009 – Jessie J. Merkay, PhD

20. https://hareesh.org/blog/2016/2/5/the-real-story-on-the-chakras

About the Author

BHAVESH C. BHAGAT

(VIRGINIA, VARANASI, AND VADODARA)

HTTP://LINKEDIN.COM/IN/BHAVESHBHAGAT
Twitter: @bbhagat (Yogi Entrepreneur)
Facebook: @PakhawajMridang
Facebook: @UniversalPilgrim
Facebook: @YogiEntreprenur

Author Bhavesh C. Bhagat is a Father, Entrepreneur, Independent Board Member, Electronics Engineer, musician, and Martial Artist. Above all, he is a "Universal Pilgrim" always seeking new inner and outer destinations to explore. Gratefully observing and continuously learning from the many varied experiences of Life, Places and the "Journey" itself. He has extensively traveled to many remote parts of the world and explored all continents of the world as a successful Entrepreneur, Chairman, and CEO as a Qualified Technology Executive for the Boards of Directors.

Now in his second innings of life, he is on an *"Inner Pilgrimage"* dedicating it to exploring the innermost hidden destinations of "Swara" and "Laya" in the journey of music. This second phase Journey in its fullness is to explore the spiritual dimension of Sound and Devotion. He is a student and practitioner of the Shuddhadvaita Branch of Hindu Vedanta philosophy (Vallabh Vedantin Pushti Sampradaya). He was honored to represent this Hindu philosophical school of thought in the 2018 Parliament of World Religions held in Toronto, Canada.

In his professional life, Bhavesh Bhagat is an Independent Board Member & Chairman of several global organizations. He is an International Public Speaker, Author & Expert on subjects of Cyber Security, Emerging Technologies, Governance & Risk, Environmental and Health and Sustainability. He is a Serial Entrepreneur and Inventor of Governance as a Service® cloud-computing platform. For Board of Director positions, he is a Qualified Technology Executive with two decades of experience as a global Chief Executive, Board Member & Advisor to various Profit and Non-Profit Boards in diverse Industries. He now advises CEO's and Boards on Creative and Agile form of Leadership leveraging diverse human experiences of music and Martial Arts. Bhavesh mentors other Startups and advises them on how to create good Cyber Security based businesses as a Board Level Mentor at Mach37 Incubator in Virginia.

He is a lifelong student and practitioner of Applied Philosophy in the form of Indian Classical music practicing oldest Indian Percussion Instrument Mridang (Pakhawaj). As a student of Vocal singing, he is focused on the practice of the Brij Bhasha Padas sung in Dhrupad and Khayal styles. He is also a student of the Indian Bamboo Flute (Bansuri). In this way, Bhavesh is striving in the present second innings of life to initiate the musical journeys on all three fronts, Taal, Swara, and Instrument. Each Taal and each Swara would take at least one lifetime to understand and that too in their basic forms, so Author has plenty of time and is in no rush to complete this musical Journey in this life.

Music for this Author is not an art, instrument or means to get "somewhere". Music for this Author is the irreplaceable joyous ingredient of offerings to the Lord.

Bhavesh is a 2nd-degree Kukkiwon Board-certified Black Belt Martial Artist in TaeKwonDo (South Korean) form of Martial Arts. The author has synthesized all his unique expertise in leading Global Organizations, music and Martial Arts to deliver and teach the Boards and CEO's the need to diversify the Leadership skills to continuously survive in face of social, technological and geopolitical winds of change we face with Agility and Creativity.

Other books by Author

Science of Rhythm - An English synopsis with some of Author's own experiences and interpretations on Science of the Rhythm. The book deals with the subject of illustrating the Science (logical and experimental roots) and Sensibilities (emotional and spiritual feeling based genesis) of the Indian Rhythmic structures also known as "Taal Shastra" in Sanskrit. The book is the first reference of its kind in the English language to explore the synthesis of Science, Spirituality, and Art in the context of Rhythmic Instruments and their structured systems of Indian Classical Musical. The first analytical and scientific section is based on the Author's continuously evolving experiments in the practice of Naad Yoga. Second, third and fourth sections go into the ancient details of the structure and science of the Taal system in India with the English descriptions and explanations of the Sanskrit terms and their meanings. The purpose of the book is to act as a reference and inspiration to educate the practitioner of Music on true science and sensibilities of the Taal Rhythmic structures so that one can perform with the full and complete depth that is gained by exploring the subject from all avenues. Available on Amazon and Public Libraries.

Also Available on Amazon Kindle Unlimited – Science of Rhythm Kindle Edition

About the Publisher

Universal Pilgrim Productions (UPP) is a global creative and media-independent publisher with a mission for unifying the scientific, artistic, philosophical and spiritual dimensions of our human existence. Having Eastern and Western readers in mind, at UPP, we believe in an infinitely joyous journey of seeking knowledge. A **"universal pilgrimage"** where the physical destination is irrelevant.

We commit our publishing projects to a Pilgrimage of human existence... of any eager soul striving to make efforts on the path of learning. We strive to publish multiple series of rare out of print and important global cultural assets in the present digital medium of Kindle and other future media as they evolve. Our mission is to benefit humanity by sharing of global cultural values and literary assets. *For future projects that fit our mission, please do contact us via email pilgrimuniverse@gmail.com*

We strive to use the latest platforms such as Amazon and other present and future media to publish at the lowest cost allowed by the platform with a predominantly non-profit mission. Where applicable all our publications depending on their nature would be disseminated free of cost using the Kindle Unlimited global platform for Kindle readers. *For all non-profit projects, we commit to our love for animals that any nominal material proceeds would be donated to the betterment of Cows in North America.*

The present volume is the new series of publications under the "Nada Yoga" series by UPP and is disseminated in the non-profit model with all proceeds donated towards Cow welfare in North America. All future volumes in this "Naad Yoga" series will have the same non-profit mission.

UPP has published several volumes on Amazon in the ongoing non-profit "Vallabh Vedant" series focusing on the most important and significant Eastern Hindu Philosophical works in Hindi, Brij Bhasha, Gujarati and English languages of Mahaprabhu Shri Vallabhacharya and his Pure Non Dualistic philosophy "PushtiMarg - Path of Divine Grace." Our humble devoted effort is to forever make these Kindle centered publications available at no cost.

Virginia, Varanasi, Vadodara

May the Light Guide the "Laya" of the Seeker
Tulsi Ghat, Varanasi, Photo by Author

A miniature from Author's personal collection depicting Kangra style of Mridang and Dance movements of Lord Shri Radhe Krishna acquired in Rajasthan, India in 2007.